Elementary
Electricity
and
Electronics
component by component

Elementary
Electricity
and
Electronics
component by component

Mannie Horowitz

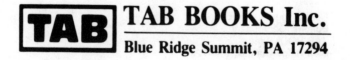

TAB BOOKS Inc.
Blue Ridge Summit, PA 17294

FIRST EDITION

THIRD PRINTING

Printed in the United States of America

Reproduction or publication of the content in any manner, without express permission of the publisher, is prohibited. No liability is assumed with respect to the use of the information herein.

Library of Congress Cataloging in Publication Data

Horowitz, Mannie.
    Elementary electricity and electronics—component
by component.

    Includes index.
    1. Electronics. I. Title.
TK7816.H73   1986   621.381   86-5961
ISBN 0-8306-0853-2
ISBN 0-8306-2753-7 (pbk.)

Front cover: reprinted from COMPUTERS AND ELECTRONICS
            copyright © 1984 Ziff-Davis Publishing.

Questions regarding the content of this book
should be addressed to:

    Reader Inquiry Branch
    Editorial Department
    TAB BOOKS Inc.
    Blue Ridge Summit, PA 17294

Contents

Introduction

The word "hobby" has many definitions. After discounting the ones about the horse and the falcon, we come to the Webster description that applies to just about every one of us. A hobby is "an occupation or interest to which one gives his spare time." It is almost always an absorbing undertaking. Considering these portrayals, electronic projects fall well within the category of being hobby-oriented exercises.

Besides being treated as a hobby, electronics should be approached as a science. A knowledge of electric and electronic components is essential if the hobbyist is to be able to undertake many different types of projects and successfully accomplish his ventures. One of the primary goals of this book is to make the hobbyist aware of the many facets of this subject without covering material meant for the designer or licensed electrician. He wants to know what a component or circuit is, how to use it, what it does, how to check it, and so on.

Technicians or would-be technicians are not ignored. Much of the material taught in schools and published in books or magazines that is aimed at the technician goes so deeply into the topics that the reader loses sight of the practical aspects of the discussion. Here the scope is limited to what the technician must know in most job situations.

There is some mathematics in the text. This cannot be avoided when discussing a science. It is assumed that you are familiar with basic algebra. Even if you do not know or cannot follow the mathematics, you can just skim through these sections of the text and use all other material to good advantage. I urge you to get involved as best you can with the mathematics detailed here.

The text starts with a discussion of what electricity is. The discussion continues by defining the different units or factors used to describe the various terms that indicate the behavior and characteristics of electricity. Details are presented about passive components such as resistors as well as about the different types of active solid-state circuit components. The discussion indicates how these components affect the various factors in electric circuits and how the various factors in the electric circuits affect the components. Because of the proliferation of digital circuits in consumer equipment and in test instruments, the last chapter is devoted to an introduction to digital logic.

In general, this book presents a practical view of current topics in electricity and electronics. After studying the text, you should have a good and relatively complete perspective on electricity and electronics in general and of the topics covered here in particular.

Chapter 1

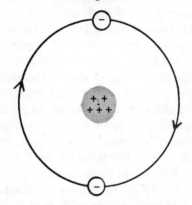

Origins of Electricity

It may seem an unlikely fact, but the light bulb is a relatively new application of electricity. Actually, electricity has been around in various naturally occurring forms for millions of years—from lightning to electric eels. Since the invention of the electric light, electricity has also found application as a source of power for radios, television sets, computers, and so on. Even though the practical applications of electricity have expanded widely within the last century, static and current electricity were known to science for many generations before that. People were studying it long before Benjamin Franklin flew his famous kite. All electricity originates from particles in the atom.

THE ATOM

All material is composed of tiny particles called *molecules*. Molecules are the smallest particles that have the characteristics of the material they comprise. A group of one type of molecules forms paper; another type forms water; a third type forms air. Regardless of whether the type of molecule being studied is used to form a solid, liquid or gas, the molecule is too small to be visible even when the material is observed through the use of the most powerful microscope. An extremely large number of one type of molecule must be present before you can see or feel the item formed by that conglomeration. Solids are formed from tightly packed groups of molecules while those forming air are very loosely packed.

1

Molecules are far from being the smallest particles of matter. Molecules are composed of smaller particles referred to as *atoms*. There are one hundred and three known types of atoms. Atoms are the basic parts of simple materials referred to as *elements*. Oxygen, hydrogen, carbon, copper, silver, gold, and so on, are examples of elements. Each of these elements is formed from a different type of atom. But only one type of atom is the basic item in each element. Molecules are usually *compounds* formed of combinations of different types of atoms. Consider water, for example. It is composed of two atoms of hydrogen and one atom of oxygen. Hydrogen has been assigned the symbol H and oxygen the symbol O, so the symbol or formula for water is the famous H_2O, indicating that there are two atoms of H in water for each atom of O. Similar formulas are used to indicate the elements and the relative quantities of each type of atom used to form other molecules.

Not all molecules consist of more than one element or one type of atom. Examples of these are hydrogen and oxygen, whose molecules are each made of two atoms. Their symbols in molecular form are H_2 and O_2. Other atoms are perfectly stable when they are separated into single atoms. These atoms may be considered either as atoms or as molecules. Copper, gold and silver are atoms which also qualify as molecules. They have been assigned the special name *monatomic molecules*.

Although the concept of the atom was known for about 2400 years before 1897, it was in that year that British physicist J. J. Thompson decided, from experimental evidence, that the atom was composed of a variety of different particles. Two of the basic particles were referred to as protons and electrons. About 1913, scientists Rutherford and Bohr proposed a model for the actual structure of an atom. It is this structure which is the basis for many of the theories of matter and electricity that we accept today.

Atomic Structure

Since the time the Bohr theory of the atom was first presented, the theory of the atomic structure has been modified and complicated to a considerable degree. Many different components were discovered in the atom. For our discussion, I need only consider three components—the proton, the electron and the neutron. The nucleus or center of the atom consists of two groups of these particles, namely protons and neutrons. The electrons revolve in specific orbits around the nucleus. These orbits may be in circular paths, eliptical paths, or paths with other shapes. In Fig. 1-1, the elec-

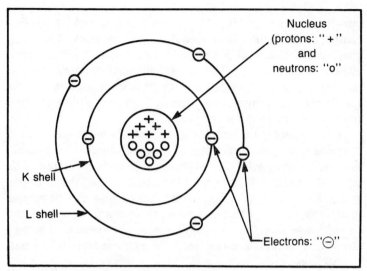

Fig. 1-1. Representative structure of an atom.

trons are drawn as being in only circular orbits.

Diverting our attention from the atom for a moment, let us turn to the bar or horse-shoe magnet. One end of the magnet is labelled the north pole and the other end the south pole. This terminology stems from what happens when a bar magnet is suspended on a string. It rotates in a horizontal path until one end of the magnet is closest to or faces the north and the other end is closest to the south. The two ends of the magnet are labelled with the names of the earth poles they face.

As far as the magnet is concerned, one end of the same piece of material is the north pole and the other end is the south pole. As for the atom, the entire group of protons and entire group of electrons have their individual types of electric charge. While the neutron is neutral and has no charge, each proton is positively charged and each electron is negatively charged. The positive and negative designations of these charges stems from the days of Benjamin Franklin and static electricity. To generate static electricity, one material is rubbed with another. When both materials attract each other or small pieces of paper, you have evidence that the materials are electrically charged. According to modern electron theory, each piece of rubbed material holds a different charge. But Ben Franklin assumed there was only one type of electric charge—a positive charge. After he rubbed two materials together, he believed that some of the positive charge had rubbed off from

3

one of the materials. Because of this assumed lack of positive charge, he referred to the material as being negative. It was discovered later that what he referred to as being negative was actually due to the presence of an excess number of electrons over the number of protons. Consequently, the individual electron became known as a negatively charged particle. The term "positive charge" remained exclusively for the proton.

Getting back to the atom in Fig. 1-1, note the negatively charged electrons revolving in orbits around the positively charged nucleus. Because the polarity of the two groups of charges differ from each other, the electron is attracted to the nucleus. But the electrons spin around the nucleus. According to physics, if a particle rotates in an orbit similar to that of a circle, there is a force to push the particle away from the nucleus. This is the phenomenon of *tangential force*. It is this force that keeps the electron from falling into the nucleus.

Two orbits or *shells* are shown in the figure, labelled K and L. An atom can have as many as seven orbits, so besides K and L, there can also be orbits M, N, O, P and Q, in that order, from the nucleus. The number of electrons that can revolve in each shell depends upon the shell involved. There can be up to two electrons in the shell closest to the nucleus, the K shell. The other shells in alphabetic order, can accommodate 8, 18, 32, 50, 72 and 98 electrons, respectively.

Different elements have different numbers of electrons in each shell. The different types of atoms have different numbers of shells. The example in the figure is that of a carbon atom. It is electrically neutral with no excess charge of either polarity, because it has the same number of protons as electrons. But the L orbit can accommodate four electrons more than shown because it is not filled to its capacity. If more electrons were added, the atom would be negatively charged. If some electrons were removed from the L shell, the atom would be positively charged.

Because specific orbits in an atom may have an eliptic rather than a round shape, electrons in shells designated by higher-order letters may be closer to the nucleus for a period of time, than are the electrons in shells specified by lower-order letters. For atoms housing orbits with this characteristic, the average distance of the electrons from the nucleus while rotating in their shell, may be less when they are in the Q orbit than when they are in one of the other orbits.

4

Valence Ring

Electrons can be added or removed relatively easily from the outermost shell of an atom. Neither protons nor neutrons are likely to be moved to or from the nucleus. If an atom has an unequal number of electrons and protons, it is no longer referred to as an atom. It becomes an ion. Should there be more electrons than protons in the atom, the ion is negative and is known as an *anion*. Should the reverse be true and the ion be positive, it is referred to as a *cation*.

There are two basic criteria which determine the ease with which electrons can be transferred to or from a particular atom. If the outermost orbit is not filled to capacity, transfer to and from that orbit can be most readily accomplished. This is especially true if lots of electrons are missing from that shell. Outermost shells in atoms are referred to as the *valence rings*. The second criterion is the distance of the valence shell from the nucleus. Electrons can more easily be moved to and from a Q valence shell than from a K valence shell, if the orbits are perfect circles. In many instances, electrons are transferred to or from the two outermost orbits, if they are only partially filled.

Materials with incomplete valence rings and a large number of orbits are generally good conductors of electricity. Electrons flow relatively easily through these conductors, for the electrons in these materials are separated rather easily from the atom. This same incomplete valence ring criterion must be satisfied if electrons are to be moved easily in and out of an atom. Table 1-1 shows a chart which indicates the number of electrons in the various shells of several types of atoms. Materials formed from these atoms are used in electric circuits.

Copper, silver, and gold have but one electron in the valence ring. But because the one electron is in the P shell for gold, in the O shell for silver and in the N shell for copper, gold should logically be the best conductor of the three elements, copper should be the worst, and silver should fall somewhere between the two. But atoms have other characteristics which indicate that we must alter our view to some extent. Because the orbits are not circular, the average distance from the nucleus of an electron in the P shell of one atom may be less than the average distance of an electron in the N shell of another atom. This occurs in the case of gold and copper. As a result, copper is a better conductor than gold. Silver is a better conductor than either copper or gold.

Table 1-1. Periodic Table of Some of the Atoms Used in Electric Circuits.

Shell	K	L	M	N	O	P	Q
Maximum Number of Electrons in Each Shell	2	8	18	32	50	72	98
Electrons in Shell of Different Elements							
Carbon	2	4					
Neon	2	8					
Aluminum	2	8	3				
Silicon	2	8	4				
Nickel	2	8	16	2			
Copper	2	8	18	1			
Germanium	2	8	18	4			
Selenium	2	8	18	6			
Silver	2	8	18	18	1		
Gold	2	8	18	32	18	1	

When considering the conducting capabilities of different metals, one additional factor should be considered. Free electrons occur in the earth and in the atmosphere. These electrons have more of a tendency to flow into some materials than into others. Materials which best accommodate these free electrons are also good conductors of electricity.

Continuing our analysis of the relative conduction capabilities of different materials, consider the characteristics of carbon, silicon, and germanium. They all have four electrons in the valence ring. Because N is its outermost shell, the best conductor of the three materials is germanium, while carbon, with four valence electrons in the L shell, is the poorest conductor of the three materials. Consequently, carbon is used in the construction of resistors. These are electric components used to retard the flow of electricity.

Germanium and silicon are used as the basic elements to form the semiconductor materials for transistors. Selenium was also popular at one time for use as a basic semiconductor material. As far as the three atoms here are concerned, the average distance of each orbit to the nucleus increases with the sequence of letters used to indicate the orbits.

Both shells of the neon atom are filled to capacity. It is a poor conductor. It will conduct only when a relatively high voltage (electric pressure) is applied to two conductors (electrodes) placed inside this gas, but separated by a minute distance. There will be an intense collision between the electrons in the atoms and the electrons present because of the applied voltage. The applied voltage forces electrons out from the neon atoms.

Whatever the atom or conducting material may be, it must be emphasized that the negatively charged electrons are the charged particles flowing through the conductor. If these electrons are forced into an atom, the atom is negatively charged. In this discussion, I will not consider the effects on the charge or, on the conduction properties of an atom when a proton is removed or added to the nucleus. As far as electric circuits are concerned, the number of protons in the nucleus never changes.

STATIC ELECTRICITY

Anyone who has ever slid out of a car during a cold and dry day and then touched the handle to close the door has experienced an electric shock. He may have even seen a spark jump from his finger tips to the handle on the door.

You have probably had the experience of walking on a wool carpet in dry weather. You then received a shock by touching a grounded piece of metal such as a water faucet, radiator, or high-fidelity set.

And there was, of course, the youthful experience of rubbing a glass rod with a piece of silk or wool, and then noting how the rod attracts tiny pieces of paper.

These are all experiences resulting from having created electricity. By rubbing one material against the other, one material loses electrons and becomes positively charged while the second material gains the number of electrons lost by the first material, thereby becoming negatively charged. For example, if you rub a rubber rod with a piece of wool, the rubber picked up a negative charge while the wool acquired a positive charge. Should you rub a glass rod with silk, the silk is negative with respect to the glass. You will find that the silk and glass are attracted to each other because their charges differ. The rubber and silk repel each other because their charges are both identically negative.

When electrons flow through an electric circuit, they flow towards the positive end of the circuit. It attracts the electrons because of the unlike polarities. These same electrons are repelled by the negative polarity at their point of origin.

One important aspect about static electricity should be emphasized. It can build up to a high voltage and yet not be harmful to the individual carrying such a charge. But certain precautions must be taken when a truck carrying a flammable liquid is involved. A chain is frequently connected to the body of the truck and trails behind it on the ground, so that the charge will be transferred through the chain to the pavement. If this is not done, a spark due to a massive discharge between the body of the truck and ground can cause the liquid to ignite.

You must, of course, also be careful not to stand under a tree when there is lightning. Lightning is an arc created by a static charge generated by air moving over the surfaces of clouds. The charged clouds discharge and arc over to the earth. If a tree is in the way, the charge will pass through the tree to the ground. If you are standing near a tree that is hit by lightning, the charge may arc over to you from the tree and flow through you to the ground. Because of the huge quantity of electric charge present in lightning, it can kill an individual standing in its path and conducting the charge.

CURRENT ELECTRICITY

When there is any flow of electrons, that flow constitutes what is referred to as current electricity. It is current electricity if it flows from the sky, through the tree and through the individual standing under the tree to the ground. Current due to a static charge is usually small and of short duration. But current used in electric circuits must be much greater and flow more consistantly than does current established by the presence of a static charge. Current must be available to flow at a constant rate through an electric circuit, and must not disappear in an instant, as it does when it is conducted from one charged item to another.

The pressure applied to electrons to flow between statically charged items is not fixed. It drops to zero at the instant the items are discharged. Where a current must keep flowing, the applied pressure between the positive and negative terminals of the source of electrons must be carefully set, controlled, and maintained.

Current

It must be remembered that electrons have a negative charge. The unit of charge is known as the *coulomb*. Each individual electron is so tiny that its negative charge is extremely small. It takes 624×10^{16} electrons to make up one coulomb of charge. One coulomb therefore represents an enormous number of electrons. But the coulomb is not an exceptionally large charge by itself. It is, however, a practical unit used to indicate the amount of charge that flows through a circuit.

When working with current electricity, charge by itself or the coulomb is seldom the factor determining the behavior of the electricity in the circuit. Charge must flow for a period of time. It is the amount of charge that flows in a specific period of time that is involved in determining the magnitude of the job done by the electricity present in the circuit. The amount of charge that passes a specific point in a circuit in one second is defined as the *current* flowing through the circuit. Electric current, rather than charge, is the practical factor considered when deliberating on items influencing the behavior of a circuit.

The *ampere* (abbreviated A) is the unit of current. It is mathematically equal to the charge divided by the time that it takes that amount of charge to flow through the circuit. Thus if 10 coulombs flow through the circuit in five seconds, there is $10 \div 5$ or 2 coulombs flowing each second. These 2 coulombs/second (the symbol "/" indicates per so that the stated number indicates 2 coulombs per second) is the number of amperes of current in the circuit.

A mechanical equivalent of electric charge or current flowing through a conductor is water flowing in a pipe. In Fig. 1-2, you can see water that may have originated in a reservoir, is trapped in a pipe. There is a cork at the end of the pipe, preventing the water from flowing out of the pipe. It just stays there and does not move. This motionless water is similar to charge staying fixed in the atom, without moving. Because the charge does not flow past a point when it is motionless, there is no current present.

Now remove the cork from the pipe for one second. Water will flow into the bucket for that second of time. A specific amount of water accumulates in the bucket during this time. This is similar to charge flowing through a wire for one second. The amount of charge due to this current is the electrical equivalent of the amount of water accumulated in the bucket during this one second. The charge-per-second is the current flowing through the wire in one second, while the accumulated water per second is the amount of water flowing through one point in the pipe in one second.

If the cork were only partially removed from the end of the pipe as in Fig. 1-2, the amount of water flowing each second would be reduced. In this position, the cork would offer resistance to the flow of water. The quantity of water flowing each second is related to the resistance presented by the cork, to the area of the pipe

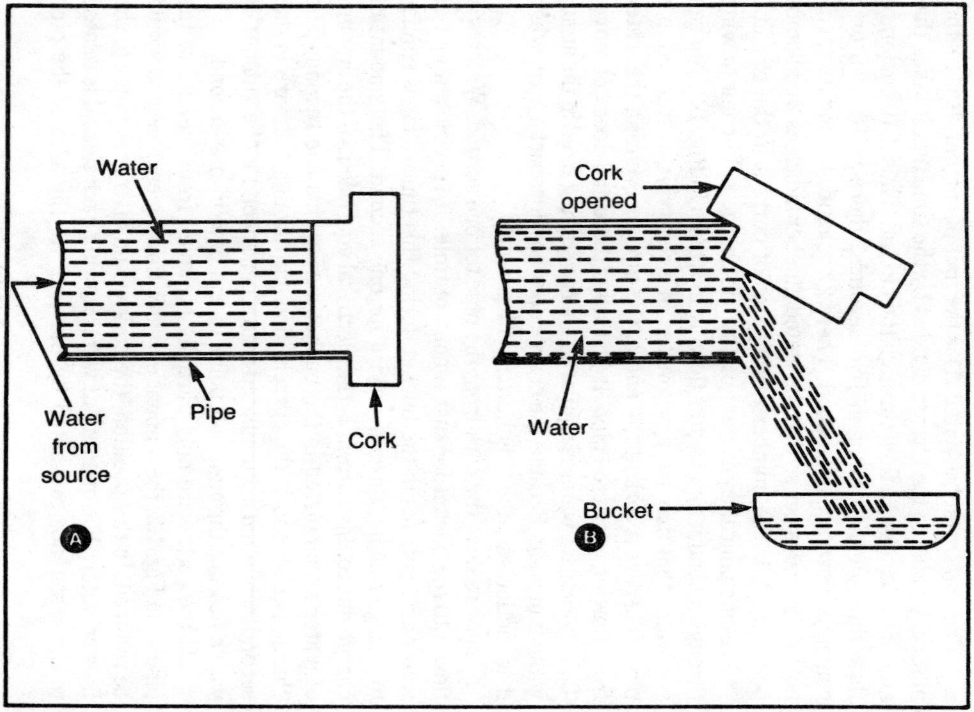

Fig. 1-2. Water flow—the mechanical equivalent of electric current.

through which the water flows, and to the pressure applied to the water by factors at the source of water or the reservoir.

Units of Current

The ampere, abbreviated as amp, or simply as A, is a very practical unit of current. For example, a 100-watt bulb connected to the standard 120-volt power line conducts somewhat under 1 ampere of current, while a 1,800 watt electric heater conducts 15 amperes of current. The ampere is a large unit of current. It is large enough to be a convenient unit for use with high-current devices.

But current in an electric circuit can be much less than 1 ampere. A battery used in a portable radio may deliver 1/100 of an ampere of current. This minute current is all that is required for the audio to have ample volume. A transistor in the same radio circuit may require about 1/10,000 of an ampere of current at one of its junctions if it is to perform its function properly. Components in the circuit affect the current level as the cork did on the water flow, i.e., by impeding it.

Fractions are awkward for handling small quantities of current. A fraction can be changed to a number in the exponential form—a very useful and convenient form when doing arithmetic manipulations. When using this form, 1/100 is equal to 10^{-2} with the minus sign (–) and the 2 being the exponents. These exponents indicate that 1 is divided by the number 10 two times. Continuing with exponential notations, 1/10,000 is equal to 10^{-4} and indicates that 1 is divided by 10 four times. But if someone should ask you how much current a battery delivers to a radio, it is awkward to say 10^{-2} amperes. Therefore, prefixes are used to compensate for these relatively clumsy verbal notations.

Where current is concerned, two prefixes are usually applied. One of these is *milli*, abbreviated m. This prefix means 1/1000 or 10^{-3}. Thus, if there is 1/1000 of an ampere of current flowing, you can just as correctly say that there is 1 milliampere (mA) of current in the circuit, or 10^{-3} amperes. A battery may supply 1/100 A of current, or 10 mA. This is derived from the fact that $10^{-2} = 10 \times 10^{-3} = 10$ milliamperes. (As you probably know, the product of two identical base numbers with exponents is that base number with an exponent equal to the sum of the original two exponents. Because $10 = 10^{+1} = 10^1$, $10 \times 10^{-3} = 10^{1-3} = 10^{-2}$.)

If there were 2/1000 of an ampere flowing in the circuit, there would be 2 mA of current. This is because $2/1000 = 2 \times 1/1000 = 2 \times 10^{-3} = 2$ mA.

The second prefix associated with the ampere is *micro*. Its abbreviation is μ, the Greek letter mu. Micro indicates 1/1,000,000 or 10^{-6} of an ampere. If there is 10^{-4} A flowing into one of the transistor terminals, the transistor conducts 100 microamperes or 100 μA of current. Should 17/1,000,000 of one ampere of 17 $\times$ 10^{-6} amperes flow in the circuit, it can be referred to as 17 μA. This follows from the fact that 17/1,000,000 = 17 $\times$ 1/1,000,000 = 17 $\times$ 10^{-6} = 17 μA.

From this discussion, it follows that 1A is identical to 1,000 mA and to 1,000,000 μA. Because the microampere is 1/1000 the size of a milliampere, it takes 1,000 μA to make 1 mA. Thus 1 mA = 1,000 μA. Other relationships following this logic are that 2A = 2,000 mA = 2,000,000 μA or 2A = 2 $\times$ 10^3 mA = 2 $\times$ 10^6 μA and that 2 μA = 0.002 mA = 0.000002 A, or 2 μA = 2 $\times$ 10^{-3} mA = 2 $\times$ 10^{-6}A.

Voltage

Current can flow through a wire or piece of electric equipment only while a pressure is applied to push the current through the wire. This is not unlike the situation where static charges of differing polarities are established on the surfaces of two different materials. Because of the presence of unlike polarities, the positive charge attracts the negative charge. This attraction establishes a pressure for the charge to travel or be pushed from one material to the other. This pressure is especially evident when there is a static charge generated due to the flow of air over a cloud. A bolt of lightning is due to the existence of an electric pressure produced by the presence of an extremely large amount of charge of one polarity on one surface and an opposing polarity on a second surface.

In an electric circuit, different methods are used to generate the pressure. This pressure is referred to as a *voltage*. One means of establishing this voltage uses a device applying chemical techniques. This device is, of course, the commonly used battery. This is also the electromagnetic piece of equipment known as the generator. In either case, the quantity of pressure is measured in units of *volts*. The symbol for voltage is V, although E is also used quite frequently to represent this pressure.

The Generator

The electric generator is based on the interrelationship between electricity and magnetism. When current flows through a wire, a magnetic field is formed near the wire. If the wire is rolled up in the form of a coil such as thread is on a spool, one end of the coil is the north pole while the second end is the south pole.

Now consider the opposite effect when a magnetic field exists because there is a permanent magnet in the vicinity, and a coil is rotated in this field. As it rotates, it breaks the magnetic field. As the field is broken, current is induced, which flows in the wires. The induced current causes a voltage to be developed between the two ends of the wire, thereby generating electricity. One end of the wire in the coil is positive while the second end is negative. Because of the differing polarities, current flows from one end or terminal of the wire to the other end when an external conductor or piece of electric equipment is connected to these ends. This current can flow through the conductor due to a pressure established across it by the different polarities existing at the two ends of the wire forming the coil.

Two opposing phenomena were just described. In one, a magnetic field was established because current was present in a wire. In the other, current was established in a wire due to the presence of a nearby magnetic field. These facts describe the operation of an electric generator. The two magnetic/electric phenomena are extremely important when they are applied to popular electric and electronic circuits and components.

The Battery

A knowledge of the chemistry of the various types of batteries in use today is not necessary to understanding the effects and performance of a particular type of battery as a voltage source. It is most important to see what a battery does and what you can expect from it.

Batteries consist of groups of individual cells. Each cell in the common dry-cell battery consists of two different metallic materials with a chemical liquid between the pieces of metal. Due to the reaction of the chemical with the metals, one metal assumes a more positive polarity than does the other.

A typical cell used in electronic equipment is shown in Fig. 1-3. The case is constructed of one type of metal, usually zinc, while a second type of metal is at the center of the cell. The type of metal used for the center terminal depends upon the type of cell formed. The case is usually negative while the center-contact is positive. Current flows through a wire or load connected from the + to the − terminal. (Note that an ordinary piece of wire should never be

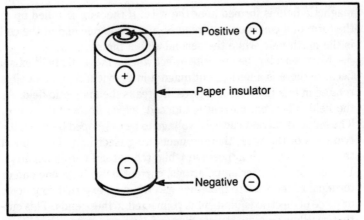

Fig. 1-3. A cell with a voltage across its terminals. Note the locations of the + and − terminals.

connected from one terminal of a battery to the second terminal. It may discharge the battery and destroy it before the battery can be used.)

The metal case of a cell is seldom exposed. If it were, there would be a good chance of the case becoming inadvertently connected to some point in the circuit where this unintentional connection is undesirable. To avoid this, a paper or plastic insulator is usually placed over the outer surface of the case. Contact can only be made with the negative metalic material at the bottom of the case.

Pressure between the voltage terminals of most cells is about 1.5 volt. Typically, lithium cells have about double this voltage while the voltage across the terminals of the mercury-oxide cell is about 10% below that of the typical 1.5 volt cell.

Carbon-zinc cells are, by far, the most popular type in use. In all cases, the larger cells can deliver more current to a circuit than smaller cells. Manufacturers of these cells recommend that the current limits for each cell not be exceeded. If they are, the life of the cell will be shortened. Current limits along with the dimensions of four popular types of carbon-zinc cells, are listed in Table 1-2.

Individual cells are frequently referred to as batteries. This is an inaccurate bit of nomenclature. Batteries are combinations of more than one cell. When cells are wired so that the positive terminal of one cell is connected to the negative terminal of the second cell, the voltage between the negative terminal of the first cell and the positive terminal of the second cell is the sum of the voltages provided by the two individual cells. This battery has 3 volts (2 × 1.5 volt) between the free terminals. If six cells are connected in this fashion, the total voltage of the combination is 6 × 1.5-volt or 9 volts. Six cells are used to form a 9-volt radio battery. The dimensions (length/width/height) of these batteries are 1″ × 5/8″ × 2″ and the recommended maximum current that should be drawn from this combination is 15 mA.

Voltage across the terminals of any carbon-zinc battery drops as current is drawn from the battery, even if it is within the specified limit. These batteries die gradually. Alkaline and mercury batteries retain their rated voltages for a long period of time, even while current is being drawn. But the voltage drops rapidly at the end of the useful life of these types of batteries. Mercury and alkaline batteries are used where the voltages must be maintained at a specific level in order to keep a particular piece of equipment operating properly.

When a battery is shown in a schematic diagram on paper, the actual physical battery is not drawn. Instead, a symbol is drawn to represent the battery. Symbols for batteries are shown in Fig. 1-4.

Start by considering the symbol of Battery #1 in the drawing. A plus sign (+) is placed next to the longer line to indicate that it is the positive terminal or pole of the battery, and a minus sign (−) is placed next to the shorter line to indicate its negative polarity. The horizontal lines connected to Battery #1 represent wires that can be used to connect the battery to other components in the circuit. The + or − symbol is frequently omitted from the schematic. If only a + is used, the − is assumed to be at the unmarked pole. If only a − is used, the unmarked pole is positive or +. If both the + and − are omitted, the longer line in the symbol is the + terminal of the battery, and the shorter line is then the − terminal.

Table 1-2. Current Ratings of Popular Carbon-Zinc Cells as Shown in Fig. 1-3.

Cell Type	Diameter (Approx.) in Inches	Height (Approx.) in Inches	Current Rating (Maximum in mA)
AAA	3/4″	1 3/4″	20 mA
AA	1/2″	2″	25 mA
C	1″	2″	80 mA
D	1 3/8″	2 3/8″	150 mA

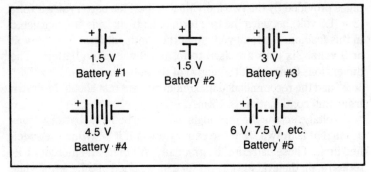

Fig. 1-4. Symbols used in schematics for cells and batteries.

Battery #1 is drawn in one position. Battery #2 is the same battery, but is drawn in another position. Only convenience in representation determines whether the battery is to be drawn vertically or horizontally. This same convenience determines whether the + or − terminal is at the left, right, top, or bottom, with respect to the other terminal of the battery.

The representation of the 3-volt battery with the two cells connected in series is shown as Battery #3 in the figure. The minus (−) terminal of one cell is connected to the plus (+) terminal of the next cell. Voltages of the two individual cells add. Voltage across the two unused terminals (one on each cell) is equal to the sum of the voltages at the terminals of each individual cell.

Two 1.5 volt cells in series add up to 3 volts. If the cells were connected with the + of one cell to the + of the second cell (or the − of one cell to the − of the second cell) as drawn in Fig. 1-5, the output from the combination would be 0 volt. The two identical voltages add, but because of the relative polarity of the two sets of terminal, the sum of the voltages is equal to zero.

When three cells are connected in series, a 4.5-volt battery is

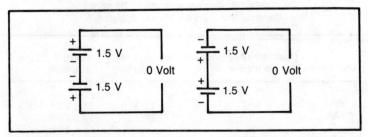

Fig. 1-5. Cells connected so that their two voltages oppose each other and add to zero.

established, as shown in the drawing of Battery #4 in Fig. 1-4. Should more than three cells be used, the symbol usually consists of several dashes drawn between the first and last cell, with the total voltage written next to the combination. This indicates the total battery voltage of the combination, although several cells were omitted from the drawing. It is drawn as Battery #5.

It was indicated above that almost any type of single cell, whatever the physical size may be, will provide about 1.5 volt. Any number of cells connected in parallel behave as a single cell, providing but 1.5 volt. As shown in Fig. 1-6, the term "parallel combination" implies that batteries or cells with identical voltages are connected across each other. Here, the − terminal of one cell is connected to the − terminal of the second cell, while the remaining + terminals are also connected to each other. Two cells in parallel are capable of providing double the current available from only one of these cells.

One type of connection between cells must never be made. Never connect the positive terminal of one cell to the negative of the other cell, while the remaining terminals of the cells are also connected to each other. This arrangement will "short" both cells and damage them irreparably. This damaging connection, involving two or more cells, is shown in Fig. 1-7.

One characteristic of the carbon-zinc cell must never be overlooked. Even when not being used, they deteriorate with time. They may leak and damage the equipment in which they are mounted. If equipment is not to be used for an extended period of time, remove the cells or batteries. Even if the equipment is being used, check the batteries from time to time to be certain that they are not leaking.

Units of Voltage

Prefixes used to indicate different categories of current mag-

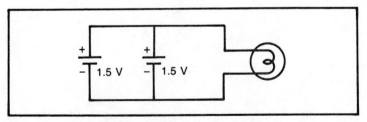

Fig. 1-6. Two cells connected in parallel. Double the current is available from two cells than is available from only one cell.

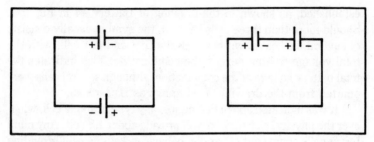

Fig. 1-7. Forbidden connection.

nitudes are used when noting the different magnitudes of voltage.

A microvolt, abbreviated μV, indicates 1/1,000,000 of a volt, just as a microampere indicates 1/1,000,000 of an ampere of current. In a similar manner, 1 millivolt (mV) denotes 1/1000 of a volt.

Although current can be in the high ampere ranges, it seldom reaches these levels. In most electric and electronic equipment, current is considerably below 1000 amperes, for that is an extremely large number. But a voltage above the 1000-volt level is quite common, although it may be a killer if you make physical contact with it. The 1000 or 10^3 (10 multiplied by itself three times) magnitude as applied to voltage, has been assigned the prefix *kilo* with the symbol k. Thus, 1000 volts is 1 kilovolt or 1 kV. Voltage applied to the picture tube in a television set can be quite high—35,000 volts or 35 kV is not an unusual potential (potential is another word meaning voltage) in these applications.

18

Chapter 2

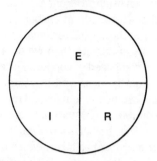

Relationship Between
Current and Voltage

Current flows in a circuit because there is a pressure or voltage pushing the negative electrons through that circuit. Voltage is one factor which determines the quantity of electrons or the size of the current flowing in a circuit. In the ideal situation—when there is a perfect conductor connected between the two terminals of the power supply—the current in the circuit is infinite because the conductor offers no resistance to the flow of current. But this is never the actual situation. There is always some resistance presented to the flow of current by the conductor. Items such as electric lights, toasters, and so on also offer resistance to the flow. In fact, every item that uses electricity has internal factors that resist the current to some degree. It is this built-in resistance that takes on special forms in different types of equipment, and which determines the function that the specific electrically driven item performs.

Complex circuits in radios, television sets, computers, and in every other type of electronic equipment use special passive components referred to as *resistors*. (A *passive* component is one that does not provide amplification while an *active* one does. An active component may be capable of providing switching action in addition to or instead of amplification.) These items restrict the flow of current in the circuits. Circuits perform their functions only because the proper resistors are chosen to do the job. The amount of current flowing in a resistor depends upon the voltage across that component as well as upon the magnitude of its resistance. This relationship is referred to as *Ohm's Law*.

RESISTORS

Different materials offer different amounts of resistance to the flow of current. If a wire made of carbon is connected from one terminal of a battery to the second terminal, it would offer lots more resistance to the flow of current than would a wire with identical physical dimensions composed of copper. The amount of resistance presented by any item is measured in *ohms*.

Electrical items can be made using a large variety of different resistances. For example, a 100-watt bulb has a resistance of about 144 ohms while a 10-watt bulb has a resistance of about 1440 ohms. A 1200-watt electric heater has a resistance of approximately 9.6 ohms while a 2-watt electric shaver has a resistance in the vicinity of 7200 ohms. In electronic circuits, resistors used can range from as low as 1/10 ohm to as high as many millions of ohms. Because of this, prefixes are used to denote the dimensions of resistors, just as they were used to help indicate the different magnitudes of current and voltage.

Units of Resistance

The basic term, *ohm*, is used as is, to indicate the magnitude of a resistance between 1 ohm and 999 ohms. Even though this statement may sound final, don't accept it as an unimpeachable rule.

The prefix kilo meaning 1000, is used when describing resistance between 1000 and 999,999 ohms. Thus a 1000-ohm resistor is usually referred to as a 1-kilohm resistor. A 12,000-ohm resistor is a 12-kilohm resistor. The abbreviation and symbol for kilohm are kohm and kΩ, respectively. Ω is the Greek capitol letter "omega." It is the symbol of the ohm.

Many pieces of digital test equipment have been designed to measure low resistances in kilohms rather than in ohms. In these instruments, the 1 through 999 ohm resistances are indicated as having 0.001 kohms to 0.999 kohms resistance. To determine the resistance in ohms after having gotten this readout, the number must be multiplied by 1000 or 10^3. Despite the kohms readout, these resistances are still referred to in the literature as being 1-ohm to 999-ohm components.

To take this one step further, resistors above 1,000 ohms are not always referred to in the literature as being in the kilohm range. They may be considered as just having a large number of ohms. Thus a 12-kohm resistor is frequently referred to as a 12,000-ohm resistor while an 860-kohm resistor is talked about as being an

860,000 ohm component.

Mega is the prefix indicating 1,000,000. Applying this to resistance, 1 megohm is 1,000,000 ohms of resistance, 22 megohms is 22,000,000 ohms of resistance, and so on. The abbreviation and symbol for megohm are Mohm and MΩ, respectively.

Just as low resistances may be represented as numbers in kilohms, resistors in the kilohm range may be expressed as decimal numbers in the megohm range. Thus 01. Mohm is 100 kohms or 100,000 ohms and 0.022 Mohms is 22 kohms or 22,000 ohms. To convert from Mohms to kohms, multiply the resistance by 1000. To convert from Mohms to ohms, multiply the number by 1,000,000.

When using these resistance values in equations to determine the voltage or current in a circuit, it is convenient to express the numbers in exponential form. The prefix kilo when expressed exponentially is 10^3. It is the same as saying 10 multiplied by itself three times, or 1000. In exponential form mega is 10^6. This is the same as saying "multiply 10 by itself six times." A 22-megohm resistor is identical to a 22×10^6-ohm resistor or to a resistance of 22,000,000 ohms.

The Shape and Size of Resistors

Resistors used in electric and electronic circuits are constructed in many different shapes and sizes. Three of these are shown in Fig. 2-1 along with how a resistor looks when it is drawn in a schematic.

The first resistor in the drawing is the most-commonly used type in electronic circuits. It is tiny so that it has the ability to conduct only small amounts of current without burning up. Its resistance can vary from 1 ohm to many megohms.

Although one drawing is presented to depict the tiny resistor, many different versions with this appearance exist. Each group has different characteristics. Of all types, the carbon composition construction is the least expensive and most frequently used. It is a mixture of fine carbon particles with a nonmetalic material. The two items forming the resistor do not fuse or combine with each other. The relative quantities of the two materials determines the resistance of the components. The resistance of components formed from this combination varies to a considerable degree with the temperature at which the resistor is being used. Resistance has been seen to change value simply by the application of heat from a soldering iron to the resistor's connecting leads. The copper leads con-

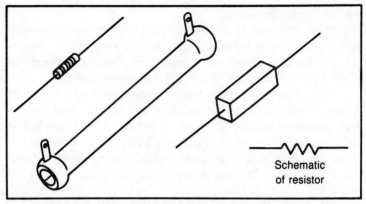

Fig. 2-1. Three shapes of resistors along with their schematic symbol.

duct the heat from the iron to the resistor.

Similar in shape to the carbon composition components are the carbon deposited and carbon film types of resistors. Here a carbon film is used to coat a ceramic rod. Unlike the carbon composition resistor, the carbon film type varies little with any temperature changes.

Thin wire has a high resistance. This is because this wire lets less current pass through it each second, than does thick wire. There is effectively less room for the electrons to flow in the thinner wire. If a long length of wire is used, it can serve as a resistor to impede the flow of current. This is the principle behind the remaining two resistors shown in Fig. 2-1.

In both cases, wire is wound around an insulated rod or a circular cylindrical form. A long length of wire can therefore be wound around a relatively short length of insulating material. The length of wire used presents a specific resistance across the terminals to which it is connected. An insulting coating is then put around the wires. The coating around the third resistor in the figure, is formed in a rectangular shape.

If the resistances of all resistors in Fig. 2-1 are identical, more current can flow safely through the wire-wound resistor than through any of the carbon types. As for the two wire-wound types, the physically larger resistor can usually accommodate a greater quantity of current.

Standard Resistor Values

Wirewound resistors can frequently be found in many different sizes and resistance values. Many manufacturers supply only those resistors shown in Table 2-1.

As is the case with all resistors, the value supplied and usually is stamped onto the component is the nominal resistance. All resistors can vary from any one of these stated values and still be within acceptable limits. This variation is known as *tolerance*. If a resistor is rated as a 10% component, it can vary from the stated value by plus or minus (±) 10%, and still be within its specified and proper value.

Consider the 100-ohm resistor. If it is a 10% component, it can vary from 10% below 100 ohms to 10% above 100 ohms and still be within its acceptable limits. Ten percent of 100 ohms is 10 ohms. Therefore, the resistor is within tolerance if its resistance is anywhere between 90 ohms (100 − 10) and 110 ohms (100 + 10). Should a 50,000-ohm resistor have a 20% tolerance specification, it can vary from 40,000 ohms to 60,000 ohms because 20% of 50,000 ohms is 10,000 ohms.

Carbon resistors are readily available in 20%, 10%, and 5% tolerances. Resistors with a tolerance of better than 5%, are frequently supplied by different manufacturers. High tolerance resistors are available only with the standard resistor magnitudes agreed on by the industry. Different values of low tolerance resistors are made available to the user by various manufacturers. There are no industry-wide standards for these components.

Standard values for 20% resistors between 1 and 6.8 ohms are shown in Table 2-2. Resistors up to 68 megohms are available in the 20% tolerance category. The standard 20% resistors are made in resistances listed in the table and multiplied by 10 for values between 10 ohms and 68 ohms, multiplied by 100 for values between 100 ohms and 680 ohms and so on, until the numbers are multiplied by 10,000,000 to indicate the standard resistors that are avail-

1	75	2,500
2	100	3,000
3	150	4,000
5	175	5,000
7.5	200	7,000
10	250	7,500
15	500	10,000
20	750	20,000
25	1000	50,000
50	1500	100,000

Table 2-1. Resistances in Ohms of Wire-Wound Resistors Usually Supplied by Manufacturers.

| 1 |
| 1.5 |
| 2.2 |
| 3.3 |
| 4.7 |
| 6.8 |

Table 2-2. Standard Carbon Resistors, 20% Tolerance, Supplied with Values between 1 Ohm and 6.8 Ohms.

able between 10 megohms and 68 megohms. Only resistors shown on the chart, or the product of these values with multiples of 10, are standard and readily available as 20% components.

A similar chart can be arranged for carbon resistors with 10% tolerance. This chart is shown as Table 2-3. Multiples of 10 apply here just as for resistors listed in Table 2-2. Ten percent resistors between 1 ohm and 82 megohms can be purchased from standard stock in values equal to multiples of 10 with the numbers shown.

Resistors with 5% tolerances are available from 1 ohm to 91 megohms. The chart in Table 2-4 shows the basic standard values. Multiples of 10 with these numbers apply here as well.

Color Code

The resistance of a large wire-wound power resistor is usually printed on the component. It may also be printed on the carbon-types of resistors. But a color code is most frequently used to identify the resistance of the carbon component, along with its tolerance. Because there are ten digits from zero to nine, there are ten colors used in this code. Each color represents a different digit. There are also two additional colors to indicate tolerance. This code is shown in Table 2-5.

Each resistor with 5%, 10% and 20% tolerance, usually has three or four color bars on its body. Resistors with closer tolerances may have either four or five color bars. Resistors falling into both groups with different numbers of bars are shown in Fig. 2-2. The bars are numbered in sequence, with the bar closest to one edge of the resistor being assigned the number 1. Some examples will

Table 2-3. Standard Carbon Resistors, 10% Tolerance, Supplied with Values between 1 Ohm and 8.2 Ohms.

1	3.3
1.2	3.9
1.5	4.7
1.8	5.6
2.2	6.8
2.7	8.2

1	1.8	3.3	5.6
1.1	2.0	3.6	6.2
1.2	2.2	3.9	6.8
1.3	2.4	4.3	7.5
1.5	2.7	4.7	8.2
1.6	3.0	5.1	9.1

Table 2-4. Standard Carbon Resistors, 5% Tolerance, Supplied with Values between 1 Ohm and 9.1 Ohms.

show just how the code is applied. I will start by considering the bar arrangement in Fig. 2-2A.

Assume you are looking for a 5% resistor with a resistance of 240,000 ohms. The first digit of the resistance is a 2, so that the bar marked "1," according to Table 2-5, should be red. Digit #2 in the number is a 4, so according to Table 2-5, the second bar should be yellow. Bar #3 is determined by the number of 0's remaining in the number after the first two digits have been dropped from consideration. For the number 240,000 there are four 0's after the first two digits. Thus the color bar marked #3 must be made yellow to indicate the four 0's after the two significant figures. Color bar #4 is used to indicate the tolerance of the component. To denote that this is a 5% resistor, that bar must be gold.

In another example, assume you have a resistor with only three of the four bars shown in Fig. 2-2A. Suppose these bars go yellow-violet-orange. Because there are only three bars, you know that this is a 20% resistor. From the color code, you also know that its value is 47,000 ohms because the yellow indicates 4, violet indi-

Table 2-5. Resistor Color Code.

Color	Digit	Tolerance
None	- - -	20%
Black	0	- - -
Brown	1	1%
Red	2	2%
Orange	3	- - -
Yellow	4	- - -
Green	5	- - -
Blue	6	- - -
Violet	7	- - -
Gray	8	- - -
White	9	- - -
Silver	× 0.01	10%
Gold	× 0.1	5%

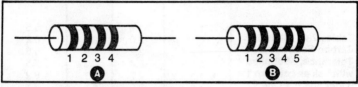

Fig. 2-2. Carbon resistors with color bars.

cates 7, and the orange indicates that there are three 0's after these first two digits.

Note that there is no tolerance bar when 20% resistors are coded. Thus there are but three color bars on a resistor with this tolerance.

As for the five-band resistors, the first three bars are for the first three digits of the resistance, the fourth bar represents the 0's remaining after the first *three* digits have been dropped from consideration, and the fifth bar indicates the tolerance. In this case, a 240,000 ohm, 1% resistor would be coded red-yellow-black-orange-brown.

A few examples should clarify any remaining difficulties with this code. Just convert from the color code to the resistance and then determine the tolerance. The solutions for various conversions are shown at the right and next to the code.

brown-red-green-silver	1,200,000 ohms, 10%
gray-red-orange-gold	82,000 ohms, 5%
orange-orange-black	33 ohms, 20%
(note that black means no zeros)	
red-red-red-red-red	22,200 ohms, 2%

Now what happens if the third bar in Fig. 2-2A is gold? That does not represent a number of 0's. But it does tell you that the resistance of the component is between 1 and 9.9 ohms. In other words, you write down the numbers in the first two digits and multiply them by 0.1 or divide them by 10. So that if the code is brown-red-gold-gold, the bars are on a 5% resistor whose value is 12/10 (12 × 0.1) or 1.2 ohms. Similarly, if the third band is silver, it means "divide by 100." If the code is red-violet-silver-gold, the resistor has a 5% tolerance and is 27/100 (27 × 0.01) or 0.27 ohms. This should not only tell you how the code works, but that resistors can have magnitudes to as low as 1/10 of those shown in Tables 2-2 through 2-4. The lowest standard resistor of any tolerance, is 0.1 ohm. Resistors under 1 ohm are seldom used and are thus not even considered when discussing the range of standard resistances. A resistor in this category usually consists of a length of bare wire with its resistance being related to the length and diameter of that wire.

OHM'S LAW

An electric circuit usually consists of a source of voltage with a resistor or other type of load connected from one terminal of the voltage source to its second terminal. The load on the supply can be any piece of electric or electronic equipment. The load is not limited to being a resistor. But where the supply is a battery or any other dc (constant voltage) source, all loads can be treated as if they are ordinary resistors. The standard values for resistors listed above obviously apply only to physical resistors. The resistances of other types of loads are not restricted to these magnitudes.

Circuit with an Ideal Battery

A circuit with a battery voltage source and a resistor is shown in Fig. 2-3. Because the − terminal of the battery is connected to its + terminal through the resistor, current flows through the resistor load. If there were any break in the circuit, no current could flow, for there would be no complete circuit to attract electrons from the negative terminal of the battery to its positive terminal. The quantity of current flowing depends upon the voltage and upon the resistance in the circuit. Assuming a perfect battery supplying a voltage E, and a resistor of R ohms at its terminals, the current in the circuit is

$$I = \frac{E}{R} \qquad \text{Eq. 2-1}$$

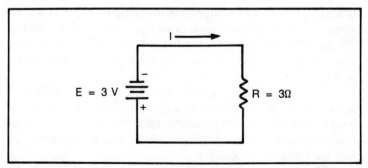

Fig. 2-3. Circuit with an ideal battery as a voltage and current source.

This is Ohm's law. Current is proportional to the applied voltage and is inversely proportional to the resistance in the circuit. Current drops as the resistance is increased.

In Fig. 2-3, if E were equal to 3 volts and R were equal to 3 ohms, the current in the circuit is 3 volts/3 ohms = 1 ampere.

Equation 2-1 can be converted algebraically to different forms. Originally, it was assumed that the voltage and resistance in the circuit were known, so all that remained was to determine the one unknown factor, the current. Instead of these factors, let us assume you know the current that you want to flow in the circuit and that you know the voltage that is applied to the circuit. You must then use the formula in Fig. 2-2 to determine the resistor necessary for establishing this condition.

$$R = \frac{E}{I} \qquad \text{Eq. 2-2}$$

Assume that the desired current is 1 ampre and that E = 3 volts. To have this current in the circuit with the 3-volt supply, the resistor determined through use of Equation 2-2 must be equal to 3 volt/1 ampere = 3 ohms.

The only other form of the Ohm's law equation is

$$E = IR \qquad \text{Eq. 2-3}$$

Here it is assumed that a specific current is flowing through a known resistor. You want to determine the voltage required across that resistor to establish this current. If I = 1 ampere and R = 3 ohms, the voltage required across the resistor to establish this condition is 1 ampere × 3 ohms = 3 volts.

The three basic forms of Ohm's law should be memorized for they are the most used relationships for electric circuits. A few examples should help clarify how these equations are to be utilized.

As a first example involving Fig. 2-3, assume E = 9 volts and R = 18 ohms. The current, I is equal to E/R = 9/18 = 0.5 ampere or 500 mA.

Next, if the voltage source is equal to 10 kilovolts and 500 μA is the current that is desired in the circuit, Equation 2-2 must be used to determine the resistance that must be in the circuit to establish these conditions. Start the calculations by converting kV into volts and μA into amperes.

$$E = 10 \text{ kV} = 10,000 \text{ volts} = 10^4 \text{ volts}$$

and

$$I = 500\,\mu\text{A} = 500/1,000,000 \text{ amperes} = 500 \times 10^{-6} \text{ amperes}$$

so

$$R = 10^4/500 \times 10^{-6} = 10^{10}/500 = 10^7 \times 1000/500$$
$$= 2 \times 10^7 = 20 \times 10^6 = 20 \text{ megohms.}$$

This all stems from the fact that 10,000 is 10 multiplied by itself four times or 10^4, 500/1,000,000 is 500 divided by 10 multiplied by itself six times or $500/10^6$, and from the fact that shifting a number with an exponent from the denominator to the numerator (or vice versa) requires only a change of sign of the exponent. This latter operation was accomplished when $500/10^6$ was changed to the form $500 \times 10^{-6}/1$. The remainder of the calculation can be accomplished using simple arithmetic.

$10^4/10^{-6} = 10^{10}$ because the sign of the exponent changes when shifting the number with the exponent from the denominator to the numerator, and also because exponents add when the numbers are multiplied. Another way of looking at this is that $10^4/10^{-6} = 10^4 \times 10^6 = 10^{10}$ because all exponents of 10 were added when they were shifted to the numerator.

If these manipulations using exponents are not obvious to you by now, it is suggested that you review them more thoroughly by finding more details about this operation in a simple mathematics or algebra book.

As a final example, assume a current of 10 mA is flowing through a 27 kilohm resistor. Determine the voltage developed across the resistor. Noting that 10 mA = 10×10^{-3} A = 10^{-2} A, and that 27 kilohms = $27 \times 1000 = 27 \times 10^3$ ohms, the voltage developed across the resistor using Equation 2-3, is (10^{-2}) (27×10^3) = $27 \times 10^{-2} \times 10^3$ = 27×10 = 270 volts.

Circuit with a Real Battery

A battery has been described as a source of voltage. It was assumed that the resistance of the battery is 0 ohm and that it was simply a voltage source. But this is never the case. A real battery does have some resistance. This resistance increases with its age.

In Fig. 2-3, the battery was assumed to be perfect. Now let us operate in the real world and assume that the battery does have resistance. A 3-volt battery with 3/10 of an ohm resistance (0.3 ohm or 3×10^{-1} ohm) is shown in Fig. 2-4A.

The resistance in the battery and the 3-ohm load resistor are connected in what is referred to as a series circuit. All current from the battery flows through the circuit connecting its − to its + terminal. This current must obviously flow through both resistors. Therefore the equivalent of the total resistance across the battery is the sum of the two resistors in the circuit, or 3 + 0.3 = 3.3 ohms. The 3-volt supply is applied to the sum of the two resistors or to the 3.3 ohm combination. The current flowing through both resistors is found by using Equation 2-1 and is 3 volts/3.3 ohm = 1/1.1 ampere.

Of the 3 volts available from the supply, the voltage across the 3-ohm resistor, according to Equation 2-3, is equal to (1/1.1 ampere) (3 ohms) = 3/1.1 volt. Using the same equation, the voltage across the resistance in the battery is (1/1.1 ampere) (0.3 ohm) = 0.3/1.1 volt. The two voltages add up to 3/1.1 + 0.3/1.1 = 3.3/1.1 = 3 volts—which is the voltage initially supplied by the battery.

Note that in a circuit with several resistors connected in series the sum of the voltage drops across all the resistors is equal to the voltage provided by the supply, regardless of whether the resistors are the sole components in the circuit or whether the resistance inside the battery is included. This voltage can never be exceeded by the sum of the voltages across all the resistors in the circuit. The total voltage can never be less than that of the supply. The supply voltage will equal exactly the sum of the voltages in the circuit.

MEASURING ELECTRICITY

Instruments are available for measuring the magnitudes of

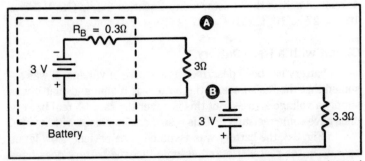

Fig. 2-4. Circuit using a real battery. This particular battery has an internal resistance of 0.3 ohm. A. Circuit with two resistors drawn separately. B. Circuit with the two series-connected resistors combined into one equivalent resistor.

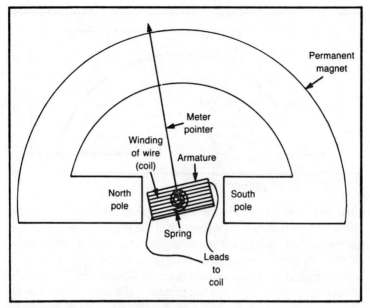

Fig. 2-5. Inside of a meter movement.

many of the active and passive components used in electric circuits, as well as to measure voltage and current. Although digital types of measuring instruments are becoming more popular and less expensive than they were in previous years, multimeters using electromechanical meter movements are still among the major pieces of test equipment on the market. They have been used by both the hobbyist and professional for many years.

The Meter Movement

The basic components of the bare instrument are shown in Fig. 2-5. A permanent magnet is mounted at a fixed location in the meter movement. A rod or armature with a coil of wire wrapped around it, is placed between the poles of the magnet. The rod is mounted in such a way that it can rotate around its center. It is kept in a fixed position by a spring.

When current is applied to the wire leads from the coil, current flows through the winding. The current establishes a magnetic field in the rod. Because of its relative magnetic polarity when compared with the polarity of the fixed magnet, the attractions and repulsions between the permanent magnet and electro magnet are such that the bar rotates. This happens because the force of the

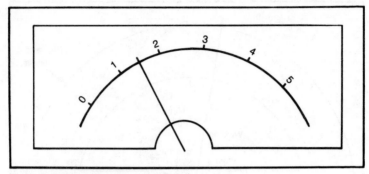

Fig. 2-6. Calibrated scale on meter panel.

field generated by the current is greater than the pull of the spring in the reverse direction. The strength of the magnetic field produced by the current flowing in the coil is proportional to the magnitude of that current. The coil rotates through a greater angle when a stronger magnetic field has been formed. Consequently, the angle of rotation depends upon the magnitude of the current fed through the winding. Angular magnitudes of rotation are directly proportional to the applied current.

A pointer is mounted on the rotating rod. As was the case with the armature, the angular deflection of the pointer is proportional to the current in the winding. A panel with a scale is usually made available. The pointer deflects to indicate a marking on the scale. This marking can indicate the current flowing through the coil of the meter movement drawn in Fig. 2-5. The scale is linear if the pointer is to indicate a number proportional to the amount of current in the coil. A panel with a linear scale and with a pointer, is shown in Fig. 2-16.

The schematic symbol for a meter movement is simply a circle with a letter printed inside the circle. Several of these symbols are shown in Fig. 2-7 along with an indication of what each symbol may represent.

The sensitivity of a meter movement is indicated by the quantity of current required to deflect the pointer to full scale. But the wire forming the coil has a resistance, R_M. The magnitude of this resistance depends upon the thickness or gauge of wire as well as upon the length of wire used to form the coil. Gauge is a standard number used to indicate the thickness of the wire. The diameter of wire represented by each gauge number is shown in Table 2-6 along with the resistance of a 1000-foot length of copper wire with that diameter. The resistance of the copper wire with a specific

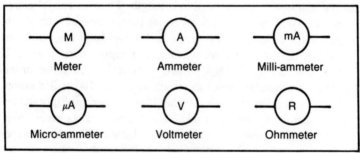

Fig. 2-7. Schematic representation of meters.

Table 2-6. Diameters and Resistance of Copper Wire.

Each factor is referred to the standard AWG number. Diameter is in inches. Resistance is in ohms for a 1000 foot length of the usual copper wire when it is used at 20° C. or 68° F. All numbers are approximate. Only even AWG numbers are shown. Odd AWG numbered wires fall about midway between the diameters and resistances shown.

AWG Number	Diameter in Inches	Resistance of 1000' Length in Ohms
0000	0.4600	0.051
00	0.3648	0.080
0	0.3249	0.102
2	0.2576	0.162
4	0.2043	0.257
6	0.1620	0.409
8	0.1285	0.650
10	0.1019	1.033
12	0.0808	1.64
14	0.0641	2.61
16	0.0508	4.16
18	0.0403	6.61
20	0.0320	10.5
22	0.0253	16.8
24	0.0201	26.6
26	0.0159	42.4
28	0.0126	67.6
30	0.0100	106
32	0.0080	168
34	0.0063	270
36	0.0050	429
38	0.0040	671
40	0.0031	1120
42	0.0025	1720
44	0.0020	2680

gauge number is proportional to is length. Thus for the 30-gauge wire, the 1000-foot length is shown as having a resistance of 106 ohms. If the length of wire is one-half of the 1000 feet shown in the table, the resistance of this length of copper wire is one-half of the 106 ohms or 53 ohms. Should the length be 3000 feet or triple that shown in the table, its resistance is 3×106 or 318 ohms. If the wire is made of material other than copper, multiply the resistance numbers shown in Table 2-6 by the number shown in Table 2-7 for the particular material involved, before calculating the resistance of the actual length of wire being used.

Getting back to the meter movement, a voltage, V_M, equal to $I_M R_M$ is developed across the coil when its resistance is R_M and I_M (current which deflects the pointer to full scale) is flowing through the winding. Thus if $R_M = 200$ ohms and 1 mA (10^{-3} amperes) flows through that meter coil, $V_M = 10^{-3} \times 200 = 0.2$ volt. The meter can thus be rated at the current or voltage required to deflect the pointer to full scale. Should 1 mA be required for a full-scale deflection of the pointer and the resistance of the movement is 200 ohms, 0.2 volt must be across the coil for the meter pointer to be deflected to full scale. When specifying a meter movement, two of these three factors must be listed in order to determine the third factor.

Table 2-7. Multiplying Factor Used to Determine the Resistance of Metal Other Than Copper For Gauges Shown in Table 2-6.

Material	Multiply Resistance in Table 2-6 by
Aluminum	1.6
Brass	4.2
Cadmium	4.4
Chromium	1.8
Gold	1.4
Iron	5.7
Lead	12.8
Nickel	5.1
Silver	0.94
Steel	10.5
Tin	6.7
Zinc	3.4

Note: With different combinations and purities, the multiplication factor for brass can vary from 3.7 to 4.9 and for steel can vary from 7.6 to 12.7.

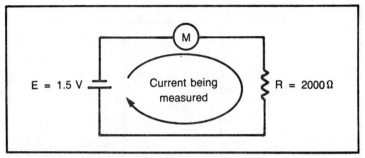

Fig. 2-8. Meter placed in series with circuit to measure current.

Current Meter

Because current flows through a circuit, the meter movement must be placed in that circuit to measure the current flowing through it. This arrangement, along with a resistor R in the circuit, is shown in Fig. 2-8. Assume that meter M is a 1 mA, 0.2 V, 200-ohm movement. The total resistance in the circuit, assuming a perfect battery, is equal to the sum of the resistance of R and the resistance of the meter movement, or 2000 + 200 = 2200 ohms. The current flowing from the battery through the meter and R, is $1.5V/2200\Omega$ or 6.28×10^{-4} A = 0.682 mA. The meter pointer deflects to indicate this current.

Since current flows through M and R, voltage is developed across these two components. The voltage across M is $200\Omega \times 0.682$ mA = 0.136 volt. Voltage across R is $2000\Omega \times 0.682$ mA = 1.364 volt. The sum of the two voltages is equal to 1.5. This total once again indicates that the sum of all voltages in a series circuit, must be equal to the voltage supplied to the overall circuit.

Voltmeters

The meter can also be used to measure voltage. The voltmeter circuit consists of a resistor wired in series with the movement. As current flows, the pointer is deflected. The angle of deflection is proportional to the current flowing through the meter due to the voltage across the resistor/meter combination. The circuit in Fig. 2-9 can be used.

As an example, assume M is a 1 mA, 200-ohm meter movement. Suppose you want the pointer to deflect full scale when 5-volts is applied to the test leads. According to Equation 2-2, this will happen when the circuit resistance is 5 volts/1 mA = 5000 ohms. Because the resistance of the meter movement is 200 ohms,

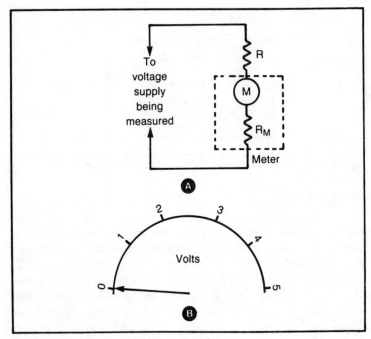

Fig. 2-9. Voltmeter circuit with 0 = 5 scale.

the resistor R in the circuit must be 5000 − 200 = 4800 ohms.

If the supply being measured is 2.5 volts, the current flowing in the circuit is 2.5 volts/5000 ohms = 0.5 mA. Deflection will be at half scale because half of the current necessary to deflect the pointer to full scale is available. Hence, when one-half of the 5 volts required to deflect the pointer to full scale is available, the pointer deflects to half scale. Should one-fifth of the voltage necessary to deflect the pointer to full scale (1 volt) be applied, one-fifth of 1 mA or 0.2 mA will flow through the meter circuit because 1 volt/5000 ohms = 0.2 mA. The pointer will now deflect to one-fifth of full scale. The entire scale is linear as far as voltage is concerned, and can be calibrated to indicate the applied voltage as shown in Fig. 2-9B.

Ohmmeter

Resistance-measuring instruments consist of a voltmeter circuit with the addition of a fixed voltage source. This is shown in Fig. 2-10. A special resistance scale decreasing in magnitude when going from left to right is also shown in the figure.

When the two test leads are connected to each other, the resistance represented by this action (between the two leads) is 0 ohm. Because of the 0 ohm resistance between the leads, the maximum current that can flow through the circuit is now available. Consequently, the pointer should be deflected to full scale by the current. For this to happen when a 1.5-volt battery supply and a 1-mA meter movement are in the circuit, the sum of R and R_M must be equal to 1.5 volt/10^{-3} amp = 1500 ohms. Since R_M = 200 ohms, R = 1300 ohms (1500Ω – 200Ω) in the particular circuit under consideration. Because full-scale deflection is the equivalent of having 0 ohm at the input, a 0 marking is placed at the maximum deflection point on the scale.

When the leads are not connected to each other, no current flows through the circuit. The resistance being measured and in series with the meter circuit is effectively infinite. The pointer does not deflect because the circuit is open and no current can flow. Thus the symbol indicating an infinite resistance, ∞, is placed at the left-hand edge of the scale.

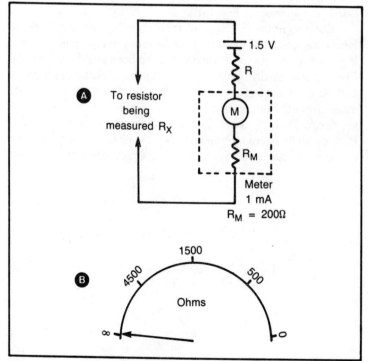

Fig. 2-10. Ohmmeter circuit with 0 = ∞ scale.

The pointer will deflect to midscale when 0.5 mA flows in the ohmmeter circuit through meter M. This occurs when the total circuit resistance, R_T, is equal to $R + R_M + R_X$. R_X is the resistance measured. Using Equation 2-2, R_T is determined as being equal to 1.5 volt/0.5 mA = 3000 ohms. Since $R + R_M$ = 1500 ohms, R_X must be 3000 ohms − 1500 ohms = 1500 ohms, for the pointer to deflect to mid-scale. Mark this resistance, 1500 ohms, at the point at the middle of the scale.

Using similar logic, the pointer will deflect to one-fourth the way up the scale when the current flow through the meter is 0.25 mA or one-fourth of 1 mA. R_T must then be equal to 1.5 volt/0.25 mA = 6000 ohms. Since $R + R_M$ = 1500 ohms, R_X must be 6000 − 1500 = 4500 ohms for the pointer to deflect to this point.

Let's calculate the resistance that will make the pointer deflect to just one more location on the scale. This time, let us find the resistor that is necessary to deflect the pointer to three-fourths of full scale to the 0.75 mA point. R_T must now be equal to 1.5 volt/0.75 mA = 2000 ohms. Because $R + R_M$ = 1500 ohms, R_X = 2000 ohms − 1500 ohms = 500 ohms. Also, this point is indicated on the scale in Fig. 2-10.

The magnitude of R_X, the resistor being measured, determines the amount the pointer is deflected when the meter is used in an ohmmeter circuit. Four points have been noted on the scale. Ohmmeters usually have more markings. Resistances causing deflections between the markings shown in the meter can be estimated from the numbers on the scale. But note that this scale is not linear. This, of course, increases the difficulty in determining R_X accurately. More intermediate points can be calculated for use on the ohmmeter scale, by employing the methods detailed above.

Chapter 3

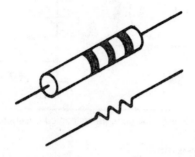

Characteristics of Resistor Circuits

Voltmeter circuits were described as consisting of two resistors—the resistance of the meter movement and the resistance of a discrete component—both connected in a series circuit. This is one of many practical applications where two or more resistors or components with resistance are connected in series.

Another practical circuit arrangement consists of several resistors connected in *parallel*. Here, two or more resistors or other electric components are connected across one another. As in a series circuit, parallel circuits are designed into many different types of equipment.

Both arrangements may be present at the same time in an individual circuit. Either or both arrangement may have been deliberately designed into some of these circuits. In other circuits, these configurations may be present although the designer did not intentionally create this array.

SERIES CIRCUITS

Two series circuits are in Fig. 3-1. In Fig. 3-1A, a battery is in series with a single resistor, R_1. In Fig. 3-1B, a battery, resistor R_1 and resistor R_2 are all connected in series. When components are connected in series, the same current must flow through all components, because there is only one path for the current.

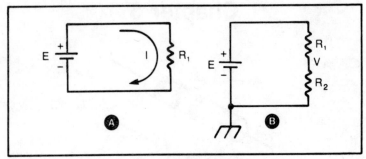

Fig. 3-1. Two circuits with components connected in series.

Equivalent Resistance

An important fact to remember is that when two or more resistors are connected in series, the equivalent resistance presented by the combination of the components is the sum of their individual resistances. If R_1 in Fig. 3-1B is 10 ohms and R_2 is 5 ohms, the total resistance supplied by the combination is $R_1 + R_2 = 15$ ohms. So if E is 1.5 volt, the current in the circuit, using Ohm's law, is 1.5 volt/15 ohms = 0.1 ampere. In A, with only R_1 in the circuit, the current would be 1.5 volt/10 ohms = 0.15 ampere.

Because the same quantity of current flows through both resistors in Fig. 3-1B, and because both components differ in resistance, a different portion of the applied 1.5 volt is across each resistor. Using Equation 2-3, the voltage across R_1 is 0.1 A × 10 ohm = 1 volt. In a similar fashion, it can be seen that the voltage across R_2 is 0.1 A × 5 ohm = 0.5 volt. Because both resistors are connected in a series circuit across the battery, the total voltage across the two resistors must be equal to that of the battery. This can be seen here when the 1 volt across the 10-ohm resistor is added to the 0.5 volt across the 5-ohm resistor. The total voltage across the two resistors is 1.5 volt.

Voltage is usually measured with respect to one point in a circuit. It is quite frequently the negative terminal of the battery or other power supply. This point in the circuit is usually referred to as the *common* or *ground*. Symbols used to indicate ground are shown in Fig. 3-2. These symbols may serve the dual function of also indicating that this point is connected to the chassis or two common land (common area) on a printed circuit board.

Voltage Divider

Utilizing the numbers determined above, assume 1.5 volt is at the plus (+) terminal of the battery with respect to its negative terminal and ground. In Fig. 3-1B, this is the same as the voltage at the upper terminal of R_1 with respect to ground. But the voltage at the junction of the two resistors and across R_2 with respect to the common ground is + 0.5 volt. Using this circuit, we can therefore establish a method of supplying a terminal with a voltage that is lower with respect to ground or the negative battery terminal than is available at the positive battery terminal.

The two resistors in Fig. 3-1B form what is known as a *voltage divider*. If the resistors are selected properly, a specific voltage less than that of the supply will be at their junction. The voltage V at the junction is related to the supply voltage E by the equation

$$V = \left(\frac{R_2}{R_1 + R_2}\right) E \qquad \textbf{Eq.(3-1)}$$

Several examples will help to indicate how the voltage at the junction varies with the resistors in the circuit. Assume E is 1.5 volt.

In the first example, let's use the resistors we mentioned above where R_1 = 10 ohms and R_2 = 5 ohms. Substituting these numbers into the equation,

$$V = \left(\frac{5}{10 + 5}\right) 1.5 = 0.5 \text{ volt}$$

This is the same number of volts that was determined by using Ohm's law.

As a second example, assume R_1 and R_2 are interchanged so that R_1 = 5 ohms and R_2 = 10 ohms. Now

$$V = \left(\frac{10}{5 + 10}\right) 1.5 = 1 \text{ volt}$$

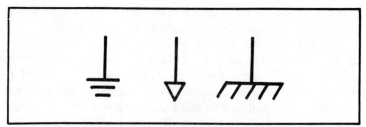

Fig. 3-2. Symbols to indicate the reference or ground point of a circuit.

If both resistors are the same, let us say 10 ohms,

$$V = \left(\frac{10}{10 + 10} \right) 1.5 = 0.75 \text{ volt}$$

Now the voltage at the junction is one-half of the voltage provided by the supply.

Let's look at all this in another way. Assume that the supply is fixed at 10 volts and that R_2 must be 20 ohms. What must R_1 be if there is 2/10 or 0.2 volt at the output?

Using Equation 3-1, $V = [20/(R_1 + 20)] 10 = 0.2$ volt. Multiply the 20 in the numerator (top number of the fraction) by 10 to get 200 while the denominator (bottom number in the fraction) remains at $R_1 + 20$. You now have the simplified equation $200/(R_1 + 20) = 0.2$. Proceeding through the steps to determine R_1, start by multiplying both sides of the equation by $R_1 + 20$.

$$(R_1 + 20)(0.2) = \left(\frac{200}{R_1 + 20} \right)(R_1 + 20)$$

so

$$(R_1 + 20)(0.2) = 200$$

so that $0.2R_1 + 4 = 200$. Solving for R_1, we find it to be equal to $(200 - 4)/0.2 = 980$ ohms.

Multirange Voltmeter

Circuits that consist of more than two resistors connected in series are quite common. One such arrangement is shown in Fig. 3-3 where there are two resistors and the resistance of a meter movement in one series circuit. A switch has been added to the circuit to short R_2 when it is closed or set in the ON position. Assume a 1 mA meter movement is being used with a coil resistance, R_M, of 200 ohms, and its scale calibrated from 0 to 5.

This is essentially the voltmeter circuit in Fig. 2-7 with the addition of a second resistor, R_2, and a switch. Assume you want to measure up to 5 volts on one range. All that need be done to establish this range is to set the switch to ON to short R_2 and make the sum $R_1 + R_M$ equal to 5000 ohms because it is equal to (5 volts/1 mA). Because $R_M = 200$ ohms, R_1 must be 5000 ohms − 200 ohms = 4800 ohms.

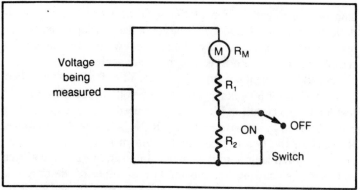

Fig. 3-3. Circuit with three resistors connected in series.

But now let us carry this a little further to a meter capable of measuring up to 50 volts. Anything above 5 volts at the input to the circuit as described, would cause more than 1 mA to flow through the meter. This would deflect the pointer off the scale and above its highest 5 volt marking. The new goal is to have the pointer deflect to 5 when 50 volts is applied to the circuit. In this way, the numbers on the meter face used for the 5-volt range can be multiplied by 10 to determine the newly applied unknown voltage. If the current is to be 1 mA when 50 volts is across the meter circuit, the total circuit resistance as determined by Equation 2-2 must be

$$\frac{50 \text{ volts}}{10^{-3} \text{ A}} = 50 \times 10^3$$
$$= 50,000 \text{ ohms.}$$

This can be achieved by setting the switch in the circuit in Fig. 3-3 to OFF so that R_2 is now in the circuit and connected in series with R_M and R_1. The total circuit resistance is now $R_M + R_1 + R_2$. Because $R_M + R_1$ is 5000 ohms, R_2 must be $50,000 - 5000 = 45,000$ ohms.

You now have a voltmeter with two ranges—5 volts and 50-volts. When the switch is closed (set to ON, 5 volts must be applied to the circuit for the pointer to deflect to 5 or full scale. When the switch is open (set to OFF), 50 volts must be applied to the circuit for the pointer to deflect to 5. The two ranges are useful. If only the 50-volt range were available, 5 volts or less would cause the pointer to deflect to the lower 1/10 , or below 10% full scale. Readings of low voltages could not be made with the same precision on the lower portion of the scale as when the pointer is deflected to near full scale. This larger deflection capability for the low voltages, is possible only when these low voltages can deflect the pointer to indicate a reading near full scale.

Should you want a voltmeter with facilities for full scale deflection when either 50 volts, 5 volts, or 0.5 volt is applied across the test leads, the circuit in Fig. 3-4 can be used. Here, four resistors are connected in series, namely R_1, R_2, R_3 and (meter resistance) R_M. Unlike the two position switch drawn in Fig. 3-3, a three-position switch is used in this circuit. In position #1 of the switch shown in the drawing, the moving arm or wiper of the switch is connected to its terminal #1. This shorts resistors R_1 and R_2, leaving only R_3 in the circuit along with R_M. In the second switch position, the wiper is connected to terminal #2 of the switch, shorting only R_2. As for the third switch position, the wiper is connected to terminal #3. Because this terminal is not connected to any point in the circuit, no resistors are shorted when the switch is put into this setting.

In switch position #1, with only R_3 and R_M in the circuit, 1 mA must flow when 0.5 volt is applied. Therefore $R_3 + R_M$ must be equal to 0.5 volt/10^{-3}ampere = 500 ohms. Because R_M was previously indicated as being equal to 200 ohms, R_3 = 500 ohms – 200 ohms = 300 ohms. For the 5-volt range when the wiper makes contact with terminal #2 of the switch, you know from the previous discussion that the total circuit resistance must be 5000 ohms. Because there is 500 ohms already in the circuit, R_1 should be 5000 ohms – 500 – ohms = 4500 ohms. As for R_2, it should be

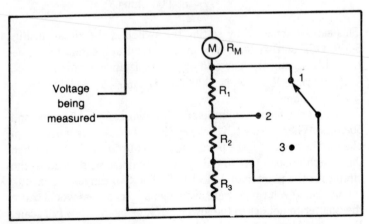

Fig. 3-4. Three-range voltmeter.

44

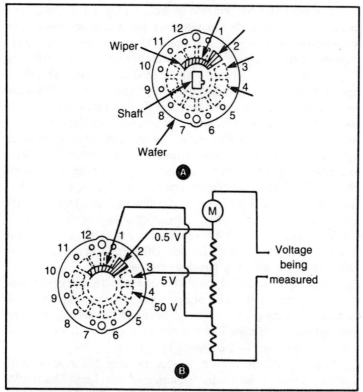

Fig. 3-5. Rotary switch. A. Designation of parts on wafer. B. Three-range voltmeter using a rotary switch to select ranges.

made equal to 45,000 ohms as before, so that $R_1 + R_2 + R_3 + R_M = 4500 + 45,000 + 300 + 200 = 50,000$ ohms, the circuit resistance required for a 50-volt range when a 1 mA meter movement is being used.

The schematic representation of the switch drawn in Fig. 3-4, can represent any one of the many different types of switches used in electronic and electric equipment. In voltmeters, the rotary-type switch in Fig. 3-5A is commonly used. It is readily available at electronic parts distributors. The mechanism consists of a shaft perpendicular to the wafer. Terminals that make contact with the wiper are mounted on the wafer. A knob is usually affixed to the shaft. The knob is used to rotate the shaft. It must be rotated to set the wiper in its different positions. Wiper contacts with the terminals, complete the circuit between the appropriate terminals on the wafer. These switch elements are shown in the drawing.

Terminal #1 is the common or stator terminal. It is always connected to the wiper. As the wiper is rotated, it completes the circuit between terminal #1 and one of the other terminals. In the position shown, terminal #1 is connected to terminal #2. After the wiper is rotated 30° or one position further in a clockwise direction, terminal #1 gets connected to terminal #3. By rotating the wafer into its third position, terminal #1 gets connected to terminal #4. Only these three positions is necessary for the three-range voltmeter circuit described here. But there is actually sufficient room to accommodate twelve terminals on this wafer. In each position, terminal #1 would be connected to one of the other terminals, so that it is connected, in turn, to terminals #2 through #12. The terminal to which it is connected depends upon how much the wiper has been rotated from the setting shown in the figure.

The circuit in Fig. 3-4 is repeated in Fig. 3-5B. But this time a practical rotating switch is used. Different voltage ranges are selected by the different positions on the switch.

Many of the practical voltmeters on the market have different circuit arrangements. In all of these no more than one resistor is in series with the meter movement in any switch setting. A different resistor is switched into the circuit in each of the ranges. A schematic showing this is shown in Fig. 3-6.

In position 1, the 200-ohm meter movement is in series with a 300-ohm resistor for a total resistance of 500 ohms. This combination establishes the 0.5 volt range for the voltmeter. Similarly, in position #2, the 4800-ohm resistor is switched into the circuit so that it is in series with the 200-ohm meter. In this way, the total circuit resistance of 5000 ohms is achieved to establish the 5-volt

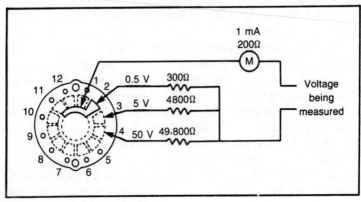

Fig. 3-6. Conventional voltmeter circuit.

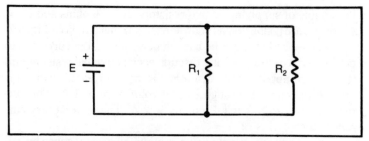

Fig. 3-7. Circuit with two resistors connected in parallel.

range. A total of 50,000 ohms, 200 ohms + 49,800 ohms is present for the 50-volt range in the third setting of the switch.

PARALLEL CIRCUITS

A circuit with two resistors connected in parallel, is in Fig. 3-7. The same voltage is across each of the resistors because both resistors are connected directly across the battery. Considering this and Equation 2-1, the current I_1 through R_1 is E/R_1 and the current I_2 through R_2 is E/R_2. The total current supplied by the battery is the current flowing through both resistors or $E/R1 + E/R_2$.

Using numbers, if E is set equal to 1.5 volt, $R_1 = 1.5$ ohms and $R_2 = 3$ ohms, the total current flowing through the circuit is 1.5 volt/1.5 ohm + 1.5 volt/3 ohms = 1 A + 0.5 A = 1.5 A. If there were one resistor, R_P, instead of two resistors, it would draw the same amount of current from the battery as is demanded by the two resistors, if R_P were made equal to 1.5 volt/1.5 amps = 1 ohm. This 1 ohm is the *equivalent resistance* of a 1.5-ohm and a 3-ohm resistor connected in parallel. The equivalent resistance of any two resistors connected in parallel can be determined by Equation 3-2

$$R_P = \frac{R_1 R_2}{R_1 + R_2} \qquad \textbf{Eq. 3-2}$$

Try it with the numbers just used where $R_1 = 1.5$ ohms and $R_2 = 3$ ohms.

$$R_P = \frac{(1.5)(3)}{1.5 + 3} = \frac{4.5}{4.5} = 1 \text{ ohm}$$

As an example, determine how much current a 9-volt battery must

47

be capable of supplying, if two resistors, R_1 = 9 ohms and R_2 = 4.5 ohms, are connected as shown in the parallel circuit in Fig. 3-7.

One method that can be used in determining the current is to calculate the current flowing through each individual resistor with the voltage applied and then add the two currents. Using this method the current through R_1 is 9 volt/9 ohm = 1 A. The current through R_2 is 9 volt/4.5 ohms = 2 A. The total battery current is then 1 A + 2 A = 3 A.

The alternate method of determining the current starts by calculating the equivalent resistance of the parallel combination. Here, it is (9 ohms) (4.5 ohms)/(9 ohms + 4.5 ohms) = 3 ohms. The current supplied by the battery to this resistance is 9 volts/3 ohms = 3 amperes. This is the same solution as when the resistors were considered individually. But the equivalent resistance method is more practical when the requirements and characteristics of different types of circuits must be determined.

Three Resistors in Parallel

It is quite simple to handle a circuit when three resistors are connected in parallel as in Fig. 3-8A. Because all three resistors are across the voltage source, E, the voltages across all of the resistors are identical. As a general rule, it should be noted that the

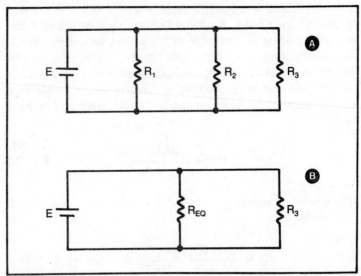

Fig. 3-8. Circuit with three resistors connected in parallel. A. Actual circuit. B. Circuit with the equivalent resistance of the R1 = R2 combination.

voltages across all resistors connected in parallel are always identical. Consequently, the total current supplied by the battery to the combination is $E/R_1 + E/R_2 + E/R_3$.

One method of determining the equivalent resistance of the three resistors connected in parallel is to consider combinations of only two of these resistors at a time. Start with R_1 and R_2. The equivalent resistance, R_{EQ} of these two resistors connected in parallel is $R_1 R_2 + R_2/(R1)$. This equivalent resistance is in parallel with R_3, as shown in Fig. 3-8B. You now effectively have two resistors left in the circuit. The parallel resistance, R_{P7} of this new combination is $R_{EQ} R_3/(R_{EQ} + R_3)$. If E is divided by R_P, the quotient is the current the battery is supplying to the resistor combination. This current is identical to that calculated when considering the sum of the currents flowing through the resistors individually.

Four or More Resistors in Parallel

The method just described for determining the equivalent resistance of three resistors connected in parallel can become very tedious if more than three resistors are in the circuit. Considering the circuit in Fig. 3-9A with five resistors in parallel, the first idea would be to keep combining two resistors at a time until the solution will finally yield only one equivalent resistance for the overall circuit. This lengthy sequence is shown in Fig. 3-9B. All steps are indicated in turn, with the equivalent circuits and appropriate formulas. The combining sequence continues until there is but one resistance, R_P, left in the circuit. R_P represents the equivalent resistance of the five-resistor combination. It is the actual resistance seen by E, the supply voltage. The number of steps in the sequence depends upon the number of resistors connected in parallel. Equation 3-3 could have been used to determine the equivalent resistance in one shot:

$$\frac{1}{R_P} + \frac{1}{R_1} + \frac{1}{R_2} + \frac{1}{R_3} + \frac{1}{R_4} + \frac{1}{R_5} \text{ etc. } [\textbf{Eq. 3-3}]$$

This formula says to calculate the quotient of one divided by each of the resistors in the parallel circuit. Then add all the numbers (each number is referred to as the reciprocal of each resistance or as the *conductance* of each resistance) that are involved. Divide this sum into 1. After going through this process, you end up with the

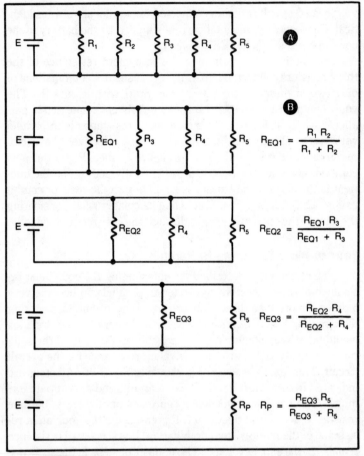

Fig. 3-9. Circuit with five resistors connected in parallel. A. Actual circuit. B. Sequence of calculations used to determine the equivalent resistance of the group of five resistors.

equivalent resistance of this combination. This is the same result that was achieved by using the procedure illustrated in Fig. 3-9B.

You now have a choice of two methods to use in determining the parallel resistance, R_p. You can take your pick. Either method supplies you with the same answer. But note that whatever R_p is, it is always less than the smallest resistance in the parallel circuit.

Current in Parallel Circuits

In an any parallel circuit, different quantities of current flow through different sections of the wire connecting the resistors. This

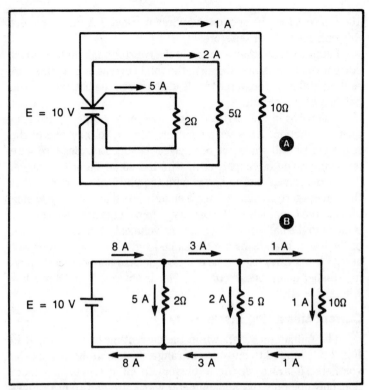

Fig. 3-10. Actual current in the leads from the battery to the resistors. A. All leads are connected directly to the battery terminals. B. Only one resistor is connected directly to the battery, while the second resistor is connected to the first resistor and the third resistor is connected to the second resistor.

can be demonstrated using the circuit in Fig. 3-10A where a power supply and three resistors are connected in parallel. Here leads from all resistors are connected directly to the battery. The total current flows only through the battery terminals. Current for each resistor flows only through the leads connected to that particular component. The total current supplied by the battery to the parallel resistor circuit, is the sum of the currents flowing through all resistor leads.

In Fig. 3-10B, the 1 A current for the 10-ohm resistor, flows through all wires. But only 1 A flows through the 10-ohm resistor and through the wires connected exclusively to that resistor.

The wires between the 2-ohm and 5-ohm resistors conduct the current for both the 5-ohm and 10-ohm resistors. Because only 2 A flows through the 5-ohm resistor and 1 A flows through the

10-ohm resistor, the sections of wires marked 3 A must conduct this current to both resistors.

Leads between the battery and the resistor bank must conduct current to all resistors. Because the total current is 8 A, this current must flow through the two leads between the battery and the balance of the circuit.

The wires in Fig. 3-10A do not perform this type of dual conduction function. Each wire conducts current to only one of the resistors. But all wires are not perfect conductors. Because wire is usually made of copper, each wire has some slight resistance. These resistances are in series with the physical resistor so that the voltage across each resistor is slightly less than the voltage supplied at the terminals of the battery. The voltage across a resistor is equal to the battery voltage less the voltage lost in the resistance of the leads. If all leads are the same length, more voltage is lost across wires carrying the larger current than across those carrying smaller quantities of current. This is obvious from Ohm's law where $E = IR$.

Current-Measuring Instruments

The milliammeter shown in the current-measuring circuit in Fig. 2-6 is limited to one current range. Just as in the case of the voltmeter, it is also very desirable for current meters to have more than only one range. A milliammeter with 1 mA and 10 mA ranges is shown in Fig. 3-11. E and R_S determine the quantity of current in the circuit.

When the switch is set to OFF R_1 is not in the circuit. Only the meter movement is in series with resistor R_S. All current flows through the movement. The pointer deflects to full scale when 1 mA is flowing.

Should R_S and E be of such magnitude that 10 mA flows through the circuit, the meter pointer will deflect off-scale and above its uppermost marking. In order to make the pointer deflect to exactly full scale when 10 mA is in the circuit, 9 of the 10 mA must be diverted from the meter movement. Then, only 1 mA is left to flow through the movement. To achieve this a shunting resistor must be connected across the meter movement as a path for the 9 mA. In Fig. 3-11, this is accomplished by setting the switch to the ON position. Now R_1 is in the circuit and shunts the 1 mA movement.

It is very easy to determine what resistor R_1 must be to do its job properly. For a full-scale deflection, the voltage across the meter

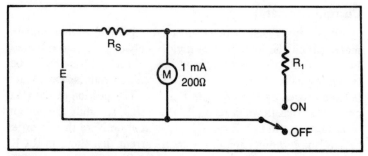

Fig. 3-11. Two-range milliameter.

movement is 1 mA × 200 ohms = 0.2 volt. Because R_1 is across the movement, 0.2 volt must be across the resistor when 10 mA is supplied to the circuit. Since 9 mA of this 10 mA flows through the resistor, R_1 must be equal to (0.2 volt)/(9 mA) ≈ 22.2 ohms. (The symbol ≈ means "*approximately equal to*.")

This idea can be expanded to a three or more range current-measuring instrument. A practical three-range circuit using a rotary switch is shown in Fig. 3-12. The ranges are 1 mA, 10 mA, and 100 mA. The resistor for use on the 100 mA range can be calculated by employing the same procedure used when setting up the 10-mA range. Now 99 mA of the 100 mA available to the instrument on this range must flow through the shunting resistor while 1 mA flows through the meter movement. The shunting resistor is 0.2 volt/99 mA ≈ 2.02 ohms.

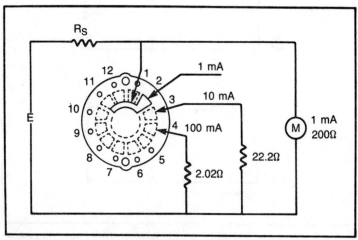

Fig. 3-12. Practical three-range milliameter circuit.

Current Divider

When two resistors are connected in parallel as in a milliam-meter circuit the total current divides between the two resistors. For the 10 mA range, the current in R_s of Fig. 3-11 is divided between the meter movement and R_1. Circuit current is also divided between resistors R_1 and R_2 in Fig. 3-7. The portion of the total current, I_T that flows in R_1 is $[R_2/(R_1 + R_2)] I_T$, while the portion of I_T flowing through R_2 is $[R_1/(R_1 + R_2)] I_T$. Note that in order to determine the portion of the total current that flows in R_1, R_2 is divided by a number equal to the sum, R_T, of the two resistors in the circuit. The portion of the current that flows in R_2 is equal to R_1 divided by R_T.

As numerical examples, I will use the circuits in Fig. 3-13. In Fig. 3-13A, R_T = 10 ohms + 5 ohms = 15 ohms. The portion of the total current that flows in the 10-ohm resistor is 5 ohms/15 ohms = 1/3, while 10/15 = 2/3 of the total current flows in the 5 ohm resistor. More current flows through the smaller resistor than through the larger resistor. This is to be expected because the voltage, E, across the two resistors is identical and I = E/10 for the 10-ohm resistor and I = E/5 for the 5-ohm resistor.

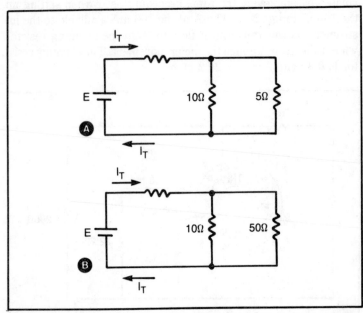

Fig. 3-13. Circuit used to determine how current is divided between parallel-connected resistors.

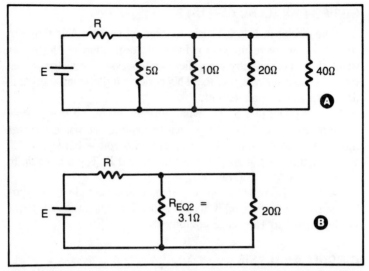

Fig. 3-14. Method used to determine the portion of the total current that flows through the 20-ohm resistor. A. Actual circuit. B. Final equivalent circuit.

Identical logic can be used for the circuit in Fig. 3-13B. Now R_T = 10 ohm + 50 ohm = 60 ohms. The portion of the total current in the 10-ohm resistor is 50/60 or 5/6 of I_T, and in the 50-ohm resistor is 10/60 or 1/6 of I_T. As always, the larger portion of current in the parallel circuit flows through the component with the lower resistance.

If more than two resistors are in a parallel circuit, other methods are available for determining the portion of the total current flowing in one of the resistors, R_x. The proper procedure is to first determine the equivalent resistance of all resistors in the parallel circuit, without involving R_x. Then use this calculated resistance as if it were a single resistor in parallel with R_x. For example, determine the portion of the total circuit current flowing through R_x in Fig. 3-14. Here, assume that the 20-ohm resistor is R_x. The equivalent resistance of the parallel circuit consisting of the 5-ohm and 10-ohm resistors is (5 × 10)/(5 + 10) = 3.33 ohms. This resistance, R_{EQ1}, in parallel with the remaining 40-ohm resistor, results in a total parallel resistance of (3.33 × 40)/(3.33 + 40) ≈ 3.1 ohms. This is R_{EQ2}. The portion of the total current in the 20-ohm resistor is 3.1/(20 + 3.1) ≈ 13/100. So if the total current from the battery is 500 mA, (13/100) × 500 = 65 mA flows into the 20-ohm resistor.

SERIES-PARALLEL CIRCUITS

The circuit in Fig. 3-11 can be used when 10 mA is required from the supply to achieve a full-scale deflection of the meter pointer. It is essentially a parallel combination of resistors in series with another resistor, R_S. This series-parallel combination is not an unusual arrangement.

If E is equal to 1.5 volt, what must R_S be for 10 mA to flow through the circuit? You know that the voltage across the meter movement and R_1 is 0.2. Consequently, 1.5 volt − 0.2 volt = 1.3 volt, remains for R_S. 10 mA will flow only if R_S is made equal to 1.3 volt/10 mA = 130 ohms.

In Fig. 3-11, you see one very elementary example of this type of relatively complex circuit. It can be found in many different forms in electrical and electronic equipment.

CIRCUIT ANALYSIS

It is frequently very enlightening to apply different theorems and rules to circuits. Through use of these methods, it is possible to analyze circuits which may, at first, appear too complex to comprehend. These processes are very useful, the mathematics and electrical concepts are simple, and the steps should become common knowledge to anyone engaged in electrical work or applying electrical concepts as a hobby.

Kirchhoff's Rules for Networks

Two Kirchhoff rules exist. They have been described or intimated in our discussions. They can be stated as follows.

1. The sum of all voltage drops in any circuit with a closed path is equal to the sum of all the voltages in that path.
2. The sum of all currents entering a particular point in a circuit is equal to the sum of all currents leaving that point.

The *voltage rule* was discussed with respect to the series circuits and expanded through use of the circuit in Fig. 3-1B. The sum of the voltages across R_1 and R_2 must be equal to the supply voltage, E. A method of analyzing a similar circuit using two batteries and two resistors in the loop can be illustrated with the help of Figs. 3-15A and 3-15B.

Analysis starts by arbitrarily selecting a direction for the current to flow through the circuit. In Fig. 3-15A, I chose a clockwise

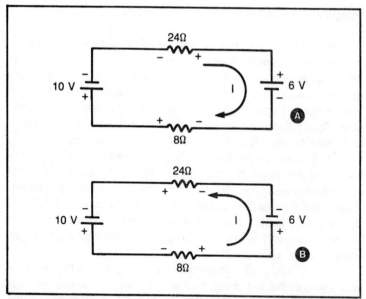

Fig. 3-15. Analysis of circuit using Kirchhoff's voltage rule.

direction. Results would be the same if a counter-clockwise direction were chosen.

When current flows through a resistor, one end becomes positive and the second end becomes negative. In Fig. 3-1A, current from E flowed through R_1. There was a voltage developed across the resistor because the current, I, flows through it. According to Equation 2-3, this voltage is IR_1. Because there is a voltage across the resistor, one end is positive and the second end is negative. Electron current flows from the negative to the positive terminal of the battery through R_1. The bottom terminal of R_1 is negative, and the top terminal is positive. This relative polarity is also true because the battery is connected with its negative terminal at the bottom end of R_1.

This fact that one terminal of each resistor must be positive or + and the second end must be negative or − is noted at the resistors in Fig. 3-15A. The end of the resistor first approached by the current is assigned a − and the second end is assigned a +. The direction of the current and polarity of the resistors will be corrected, if necessary, after the circuit has been fully analyzed.

Now you can apply three rules to determine various facts about the circuit. You want to know the current flowing in the circuit, the voltage across the resistors, and the actual polarity at the terminals of each resistor. These rules are:

1. When current flows through a battery, the battery voltage is considered positive when the arrow indicating the current is shown as entering into the positive terminal of the battery. This current flows through the battery, and originates from the negative terminal. Battery voltage is considered negative if the reverse is true and the current enters the battery in the negative terminal and flows through it to exit from the positive terminal.

2. Current enters one terminal of a resistor and exits from its second terminal. The terminal through which the current enters is considered negative. The second terminal is considered positive. This polarity is marked at the resistors in Fig. 3-15A. Just as in the case of the battery, the voltage across the resistor with the polarity shown is negative, because the current enters it through a negatively marked terminal.

3. The sum of all voltages in the circuit must add to zero. This was also true for the circuit in Fig. 3-1A where the voltage supplied by the battery and the voltage across the resistor was identical. Because of this relationship and the opposing polarities, the sum of the two voltages adds to zero.

Apply the three rules to Fig. 3-15. In this analysis, you can assume that the current can start flowing at any point in the circuit. Suppose that it starts to flow through the 24-ohm resistor. The voltage across the resistor is $-24 \times I$. Continuing in this direction of current flow, the voltage supplied by the 6-volt battery is $+6$ volts, voltage across the 8-ohm resistor is $-8 \times I$, and the voltage supplied by the 10 volt battery is $+10$ volts. Applying the third rule, the sum of all voltages in the loop is zero or

$$-24I + 6 - 8I + 10 = 0$$

Solving for I, you will find that $-32I = -16$ and that $I = 1/2$ ampere.

In Fig. 3-15B, the polarity of the 6-volt battery and the direction of current flow are reversed. Here, the voltage across the 24-ohm resistor is $-24I$, and the voltage across the 10-volt battery is -10 because the current enters it through the negative terminal. The voltage across the 8-ohm resistor is $-8I$, and the voltage across the 6-volt battery is $+6$. Combining these bits of information, the equation for the loop becomes

$$-24I - 10 - 8I + 6 = 0$$

or $-32I = +4$. Now, I is equal to $-4/32 = -1/8$ ampere. I obviously chose the wrong direction for I because the solution shows the current to be negative. The real electron current actually flows in the opposite direction. The polarity of the voltages across the two resistors are therefore opposite to those shown in the drawing.

A more involved circuit is shown in Fig. 3-16. The aim here is to find the voltage across the 24-ohm resistor.

Current I_1 flows in the left loop and I_2 flows in the right loop. Polarity of voltages across all resistors due to both currents are as shown. Note that the polarity of the voltage across the 24-ohm resistor, is in one direction due to I_1 and in the opposite direction due to I_2. Both polarities are noted at the respective sides of the resistor. The voltage across the resistor is due to the difference of the two voltages caused by the presence of both I_1 and I_2. It is effectively $24 I_1 - 24 I_2$ because the polarities oppose each other. The relative magnitudes of I_1 and I_2 determine the direction of the actual current flowing through the resistor.

Summing the voltages in the left loop,

$$+48 - 8I_1 - 24I_1 + 24I_2 = 0$$

because the voltage developed across the 24-ohm resistor due to I_2 affects I_1. This voltage is added to the other voltages in the equation because the polarity of the voltage across the resistor due to I_2 is opposite to the polarity of the voltage due to I_1. In a similar fashion, the voltages in the right-hand loop are

$$+72 - 24I_2 - 12I_2 + 24I_1 = 0$$

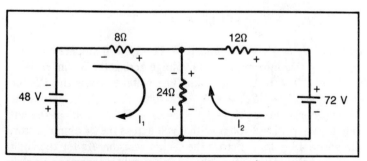

Fig. 3-16. Voltage across the 24-ohm resistor determined in circuit with two currents using Kirchhoff's rule.

Simplifying the two equations,

$$32I_1 - 24I_2 = 48$$
$$-24I_1 + 36I_2 = 72$$

Two equations arranged as shown, with the same two variables in both, are known as *simultaneous equations*. Two equations are necessary in order to find the magnitudes of two unknown currents. You can determine what one of these unknowns is by applying methods to eliminate the second unknown. The two equations are combined into one equation with one unknown. To do this, the numbers multiplied by the same variable in both equations must be made identical. The equations are then either added or subtracted from each other, as required, to eliminate one of the variables.

When the multipliers of a particular variable are equal in both equations, the difference (or sum) of the two equations will result in a zero factor where this variable was. To accomplish this for the equations shown, multiply the upper equation by 3 and the lower equation by 2. The resulting equations are

$$96I_1 - 72I_2 = 144$$
$$-48I_1 + 72I_2 = 144$$

Adding the individual terms leaves us with

$$48I_1 - 0I_2 = 288$$

so that $I_1 = 288/48 = 6$ amperes. Once you know I_1, it is substituted into either one of the original equations to determine I_2. Using the second equation,

$$-24 (6) + 36I_2 = 72$$

so that $36I_2 = 72 + 144$ and $I_2 = 216/36 = 6$ amperes.

Using this information, the voltage across the 8-ohm resistor is $8I_1 = 8\ \Omega \times 6$ A = 48 volts. The voltage across the 12-ohm resistor is $12I_2 = 12\ \Omega \times 6$ A = 72 volts. The voltage across the 24-ohm resistor is $24 (I_2 - I_1) = 24 (6 - 6) = 0$ volt. In the left loop the supply is 48 volts. With 0 volt across the 24-ohm resistor, the entire 48 volts is across the 8-ohm resistor. As for the right loop, the entire voltage from the 72-volt supply is across the 12-ohm resistor because 0 volt is across the 24-ohm resistor.

This procedure can be applied with equal success to a circuit where I_1 and I_2 are not equal. Consider the components shown in Fig. 3-17. What are the voltages across the various resistors? Polarities due to both currents are marked at all resistors. Note that polarities of voltages developed across the 10,000-ohm resistor due to I_1 and I_2 are identical, so that voltages developed by these two currents add. The equations are

$$100 - 2000I_1 - 10,000I_1 - 1000I_1 - 10,000I_2 = 0$$
$$200 - 10,000I_2 - 4000I_2 - 10,000I_1 = 0$$

Simplifying the equations,

$$13,000I_1 + 10,000I_2 = 100$$
$$10,000I_1 + 14,000I_2 = 200$$

Multiplying the top equation by 14 and the bottom equation by 10 yields

$$182,000I_1 + 140,000I_2 = 1400$$
$$100,000I_1 + 140,000I_2 = 2000$$

Subtracting the bottom equation from the top one results in the relationship $82,000I_1 = -600$ so that $I_1 = -7.3 \times 10^{-3}$ A, a negative current. Substituting this into the first original equation,

$$13,000 (-7.3 - 10^{-3}) + 10,000I_2 = 100$$

we find that $10,000I_2 = 100 + 94.9$, so that $I_2 = +19.5 \times 10^{-3}$ A.

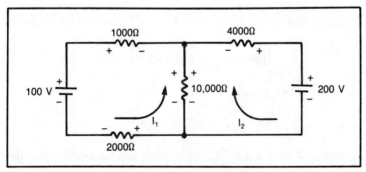

Fig. 3-17. Kirchhoff's voltage rule is applied in this second example.

Using this information, the voltage across the 1000-ohm resistor is $(-7.3 \times 10^{-3})(1000) = -7.3$, voltage across the 2000-ohm resistor is $(-7.3 \times 10^{-3})(2000) = -14.6$, voltage across the 4000-ohm resistor is $(19.5 \times 10^{-3})(4000) = 78$, and the voltage across the 10,000 ohm resistor is $(-7.3 \times 10^{-3})(10,000) + (19.5 \times 10^{-3})(10,000) = -73 + 195 = +122$ volts.

When determining the voltages across the 1000-ohm and 2000-ohm resistors, you found the voltages to be negative. Voltage across the 1000-ohm resistor is -7.3 and across the 2000-ohm resistor is -14.6. The fact that the voltages showed up in the equations as being negative, indicates that the actual polarity of voltages across these resistors are opposite to those shown in the drawing. But when added to all other voltages in the left loop, the sum must equal the 100 volts supplied. The sum of the two negative voltages in that loop is $-14.6 - 7.3 = -21.9$ or about -22 volts. The voltage across the 10,000 - ohm resistor is $+122$ volts. The total voltage in the loop is $+122 - 22 = 100$ volts. This is identical to the voltage supplied by the voltage source in this loop.

As for the loop at the right half of the drawing, when the 122 volts across the 10,000-ohm resistor is added to the $+78$ volts across the 4000-ohm resistor, the total is 200 volts. This is identical to the voltage supplied by the voltage source in that loop.

Note that the polarity of the voltages across the resistors in the left loop are the reverse of those shown. This is due to the relative magnitudes of the voltages and currents in the two loops.

As a final example, let us reverse the 200-volt battery and determine just what the various voltages and currents are in the circuit. This is shown in Fig. 3-18. The equations are

$$13,000I_1 - 10,000I_2 = 100$$
$$-10,000I_1 + 14,000I_2 = 200$$

Multiplying the top equation by 14 and the bottom by 10 yields

$$182,000I_1 - 140,000I_2 = 1400$$
$$-100,000I_1 + 140,000I_2 = 2000$$

Adding the two equations yields

$$82,000I_1 = 3400$$

so that $I_1 = 41.5 \times 10^{-3}$ A. Substituting this information into the first equation,

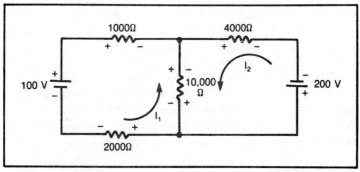

Fig. 3-18. Circuit identical to the one in Fig. 3-17, but with the connections to the 200-volt supply reversed.

$$539.5 - 10,000I_2 = 100$$

indicates that

$$I_2 = 43.9 \times 10^{-3} \text{ A.}$$

Using Equation 2-3 and the values determined for I_1 and I_2, I find that the voltage across the 1000-ohm resistor is 41.5, across the 2000-ohm resistor 83, across the 4000-ohm resistor 175.6, and across the 10,000-ohm resistor it is $(43.9 \times 10^{-3})(10,000) - (41.5 \times 10^{-3})(10,000) = 24$ volts. The polarity of the voltages across the two ends of the 10,000-ohm resistor is primarily due to the direction of I_2. This is because the voltage developed in that resistor when I_2 flows is greater than the voltage developed by the I_1 current.

Now add all voltages in the circuit. In the left loop, the total voltage is $41.5 + 83 - 24 = 100.5$. This is close to the 100 volts provided by the voltage source in that loop. In the right loop, voltages add to $175.6 + 24 = 199.6$. The difference between the calculated 199.6 volts and the actual 200 volts supplied, as well as the difference between the calculated 100.5 volts and the actual 100 volts supplied, is due to calculation tolerances. These differences can be ignored. If more significant figures were considered in the calculations, the errors would not exist.

Kirchhoff's current rule can be described using Fig. 3-17. Current I_1 flows from the 100-volt supply. Current I_2 flows from the 200-volt supply. Both currents are at one junction in the circuit, along with the current flowing in the lead connected to the 10,000-ohm resistor. According to Kirchhoff's rule, all current entering a junction must flow out of it. Since I_1 and I_2 entered the junction, $I_1 + I_2$ must flow out of it. The only place for this current to flow is through the 10,000-ohm resistor. Because $I_2 = 19.5 \times 10^{-3}$ A and $I_1 = -7.3 \times 10^{-3}$ A (both were calculated above), the total current that flows out from the junction is $19.5 \times 10^{-3} - 7.3 \times 10^{-3} = 12.2 \times 10^{-3}$ A. This current flows through the 10,000-ohm resistor. Because of this current, the voltage across the resistor is (12.2×10^{-3}) (10,000) = 122 volts. This is the same voltage calculated earlier when I applied Kirchhoff's voltage rule. It just confirms that all current entering any junction is equal to the sum of the currents flowing out of that junction.

Thevenin's Theorem

Circuit analysis can be made less complex by establishing simplified equivalent circuits for complex arrangements. Determining one equivalent resistance for a group of parallel or series-connected resistors is an example of a simplified representation of a group of resistors. Now suppose you have a complex circuit with resistors and a power supply. How would you proceed to determine the voltage and current flowing through one of the resistors? Thevenin's theorem can be applied here.

All elements in the circuit are considered when applying the Thevenin theorem, other than the load. Using methods described here, the load is removed from the circuit. Components under consideration are combined to form a circuit consisting of a voltage in series with a resistance. When the load is once again connected to this circuit, it sees this simple arrangement. Because of this simplicity, it becomes a trivial matter to calculate the load current and voltage.

Three steps are required to determine the Thevenin equivalent of any circuit.

1. Remove the output resistor or other load from the circuit.

2. Short the voltage source (on paper, of course). With this voltage source shorted, determine the total resistance seen by looking into the terminals from which the load was removed.

3. Break the short that you placed across the supply. Determine the open circuit voltage. This is the voltage at the load terminals after the load has been removed.

Using these methods, let us determine the Thevenin equivalent of the circuit in Fig. 3-19A. If the 216-ohm resistor is the load, the aim here is to determine the current flowing through and the voltage across that resistor. As a first step, remove the load from

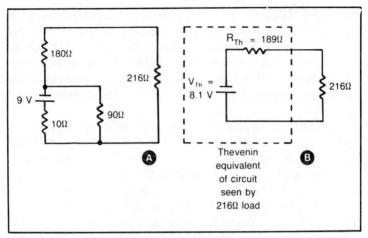

Fig. 3-19. A. The Thevenin equivalent of this circuit is to be determined. B. Actual Thevenin equivalent of circuit in A.

the circuit. Next, short the power supply. These are Steps 1 and 2 in the procedure. After this has been done, a 10-ohm resistor is in parallel with a 90-ohm resistor to form an equivalent resistance of $(10)(90)/(10 + 90) = 9$ ohms. Looking from the load terminals, this 9-ohm combination is in series with the 180-ohm resistor for a total equivalent circuit resistance of 189 ohms. The load will see this resistance after it has been reconnected to the circuit. Thus, 189 ohms is the Thevenin equivalent resistance, R_{Th}, of this circuit.

Continuing with Step 3 of the procedure, restore the circuit to the form shown in Fig. 3-19A, but leave the load disconnected. The open circuit voltage is the same at either side of the 180-ohm resistor because no current flows in the resistor when the load is removed from the circuit. Under these conditions, there is no voltage across the 180-ohm resistor for $IR = 0$ when no current flows through it. But there is always current in the 90-ohm and 10-ohm resistors, for they form a complete circuit around the battery. Because the total resistance across the battery due to these two resistors is 90 ohms + 10 ohms = 100 ohms, the open-circuit battery current using Ohm's law is 9 volts/100 ohms = 0.09 ampere. Voltage across the 10-ohm resistor is (0.09 ampere)(10 ohm) = 0.9 volt. The battery voltage, less the 0.9 volt, is supplied by the battery to the open circuit. Therefore, 8.1 volts (9 volts − 0.9 volt) is the Thevenin equivalent voltage, V_{Th}, of this circuit.

The complete Thevenin equivalent of the circuit is shown in

Fig. 3-19B. All current in the equivalent circuit flows through the 216-ohm load resistor, and is 8.1 volts/(189 + 216) ohms = 8.1 volts/405 ohms = 0.02 ampere. Due to this current, the voltage across the load resistor is (216 ohms) (0.02 ampere) = 4.32 volts. The quantity of current considered in the equivalent circuit is only the amount flowing in the load. It is not the total current supplied by the battery.

For another example of an application of Thevenin's theorem, consider the circuit in Fig. 3-20A. The aim is to find the voltage across the 30-ohm resistor and the current flowing through it.

Start by disconnecting the 30-ohm resistor from the circuit. Looking into the balance of the circuit from the disconnected terminals, the 90-ohm and 180-ohm resistors form a voltage divider. The voltage across the 90-ohm resistor is (90/270) (9 V) = 3 volts. This is the Thevenin voltage. Now short the battery. The equivalent or Thevenin resistance looking into the open 30-ohm load is 90 ohms in parallel with 180 ohms. This is equal to (90) (180)/(90 + 180) = 60 ohms. Using this data, the new Thevenin equivalent of the original circuit is shown in Fig. 3-20B. The voltage across the 30-ohm resistor using voltage divider methods is (30/90) (3 V) = 1 volt. The current in the 30-ohm resistor is 1 volt/30 ohms = 0.033 A.

Norton's Theorem

A voltage source is considered a low impedance (or resistance) supplier of voltage. The perfect source is a battery or generator without any internal resistance. With this perfect source, the same voltage would be across different loads even if their magnitudes varied considerable. But as soon as an imperfect battery or generator is used, there is an internal resistance in its equivalent circuit. Load current flows through that resistance as well as through the load. When this happens, there is voltage developed across the internal resistance of the voltage source. The voltage left for the load is the battery voltage less the voltage developed across its internal resistance. The quantity of voltage developed across the internal resistance varies with the quantity of current demanded by the load. The larger the current through the battery resistance, the larger the voltage that is developed across it. Voltage at the battery terminals is its rated voltage less the voltage developed across its internal resistance. When more current is demanded by the load, less voltage is supplied to it due to the larger voltage drop inside the voltage source.

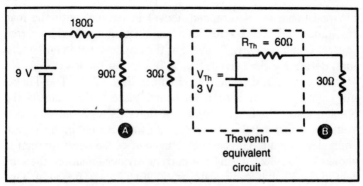

Fig. 3-20. A second example using the Thevenin equivalent of the circuit.

Now assume you have a supply that is specially designed to have a large internal resistance—let's say 1,000,000 ohms. If the supply is fixed at 10 volts and is shorted at its terminals, the current flowing into the shorted terminals is 10 volts/1,000,000 ohms = 10^{-5} amps. If the terminals are not shorted, but instead a 1000-ohm resistor is placed across them, the total resistance in the circuit and seen by the supply is 1,000,000 ohms + 1000 ohms = 1,001,000 ohms. The total current flowing from the battery is now 10 volts/1,001,000 ohms = 0.999×10^{-5} amps. If the load in a similar situation is 10,000 ohms rather than 1000 ohms, the current supplied to it is 10/1,010,000 = 0.990×10^{-5} amps. Neither of these load currents differ from the short-circuit current by any magnitude worth noting. A current source is a high-impedance (or resistance) supply providing a relatively fixed current to loads of different magnitudes. It was just shown that a supply with an internal resistance of 1,000,000 ohms is a good current source. The current supplied by any current source will change considerably from the fixed value if the load is large and comparable in size to the internal resistance of the supply. But this does not happen under ordinary circumstances.

The Thevenin equivalent circuit incorporates a voltage source—a theoretical supply with a zero internal resistance. The Norton equivalent circuit includes a current source—a theoretical supply with an infinite internal resistance. This ideal current source is shunted by the Thevenin equivalent resistance.

The steps in setting up the Norton equivalent circuit start by shorting the load resistance (on paper only). Calculate the current flowing from the supply through that short. This is the Norton current. In the drawing of the equivalent circuit, connect the Thevenin resistance (as determined above) in parallel with the load resistance and then connect this combination across the current source. Using the current source as the total current in both resistors, determine the current that is flowing in the load.

For example, consider the circuit in Fig. 3-20A. The Thevenin equivalent resistance is 60 ohms. Place a short across the 30-ohm load. The total current that flows through the short is 9 volts/180 ohms = 0.05 amp. This is the total current in the Norton equivalent circuit. The Norton equivalent of the overall circuit is shown in Fig. 3-21. Using the current divider equation, the current in the 30-ohm load resistor is (60/90) 0.05 = 0.033 amp. This is the same current that was calculated by using the Thevenin equivalent of the circuit.

As for the circuit in Fig. 3-19A, Thevenin resistance has already been determined to be 189 ohms. Next, short the load resistor. The modified equivalent circuit is shown in Fig. 3-22A. To determine the current for the Norton circuit, you must first determine the voltage lost, or the voltage drop, across the 10-ohm resistor. To do this, note that the resistance of the entire circuit as seen from the battery is 180 ohms in parallel with 90 ohms, and that combination is in series with 10 ohms. Thus, the battery sees a resistance of 10 + (180) (90)/(180 + 90) = 70 ohms. The current from the battery to this combination is 9 volts/70 ohms = 0.129 amp. This total current flows through the 10-ohm resistor so there is a 10 ohms (0.129 amp) = 1.29 volt loss across that resistor. Thus, 9 volts less 1.29 volts or 7.71 volts is across the 180-ohm resistor and short, so that the current flowing through the short is 7.71

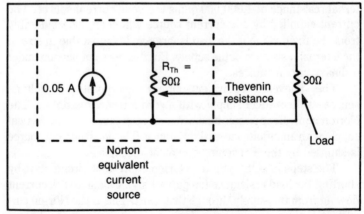

Fig. 3-21. Norton equivalent circuit of the configuration in Fig. 3-20A.

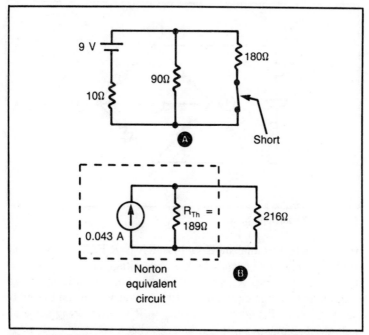

Fig. 3-22. Deriving a Norton equivalent circuit. A. Circuit with the load shorted. B. Norton equivalent of the circuit in A.

volts/180 ohms = 0.043 amp. Use this as the current supply in the Norton equivalent circuit.

With the information just determined, you can generate the Norton equivalent of the circuit. This is shown in Fig. 3-22B. Here, the 189-ohm Thevenin resistance and the 216-ohm load shunt the 0.43 amp current source. Using the current-divider equation, the current through the 216-ohm load resistor is [189/(189 + 216)] [0.043 A] = 0.02A. This is identical to the current that was determined earlier using the Thevenin equivalent circuit.

Analyzing a Bridge

The Wheatstone bridge in Fig. 3-23 is a very commonly used circuit in different devices. It is applied to measuring resistance as well as other electrical factors.

Voltage E is fed to two pair of series-connected resistors, R_1 = R_2 and R_3 = R_4. Using voltage-divider equations, the voltage between the R_1 = R_2 junction and the negative terminal of E (across R_2) is $R_2E/(R_1 + R_2)$ and between the R_3 = R_4 junction and

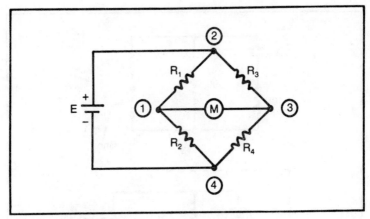

Fig. 3-23. Wheatstone bridge.

the negative terminal of E (across R_4) is $R_4E/(R_3 + R_4)$. The bridge is said to be *balanced* when the voltage across R_2 is equal to the voltage across R_4. This can be stated mathematically by setting up an equation equating the two voltages. You can therefore set $R_2E/(R_1 + R_2)$ equal to $R_4E/(R_3 + R_4)$. After doing some algebraic manipulations, it can be seen that

$$\frac{R_1}{R_2} = \frac{R_3}{R_4} \quad \text{or} \quad \frac{R_1}{R_3} = \frac{R_2}{R_4} \qquad \textbf{Eq. 3-3}$$

when the bridge is balanced.

When the voltage at the $R_1 = R_2$ junction is equal to that at the $R_3 = R_4$ junction, the voltage difference between the two points (point #1 and point #3) is obviously zero. This is no pressure to push current through the meter connecting the two junctions. So when the bridge is balanced, it can be concluded that the meter pointer does not move. It indicates 0 volt.

The relationships in Equation 3-3 indicate a method of measuring resistance through the balance of a bridge. Suppose you want to measure the resistance of an unknown component, R_x. Start by replacing R_4 in the circuit with R_x. Substitute a variable control for the fixed resistor, R_2. (A variable control is constructed of fixed resistance material with an arm or wiper moving over that resistance. The resistance between the wiper and both ends of the resistance material depends upon where the wiper rests on the fixed resistance.) Keeping the symbol R_2 for the variable control, you can use an ohmmeter to measure its resistance when the wiper is set in a specific position. Vary R_2 until the bridge is balanced and no current will flow through the meter measuring it. Substitute all resistance for the three known components into Equation 3-3 and solve for the fourth resistance, R_x. An example using this method is shown in Fig. 3-24.

When the bridge is not balanced, current I flows through the meter, M. The magnitude of this current is determined through use of the Thevenin equivalent circuit. It was noted in the previous discussion that if more current than is available from one battery is required from a supply with a specific voltage, two identical batteries can be wired in parallel. Reversing this in circuit analysis, one battery can be separated into two batteries with identical voltages as long as this does not upset the voltage relationships in the circuit. In this manner, one battery is treated as if it consisted of two batteries connected in parallel. This is shown in Fig. 3-25A for the circuit in Fig. 3-23.

You can now find the Thevenin equivalent circuit around each of the supplies. This is shown in Fig. 3-25B. The Thevenin voltage in Fig. 3-25A is that at the junction of R_1 and R_2 (across R_2) when M is disconnected from the circuit. The Thevenin voltage for the circuit in Fig. 3-25B under the same conditions is equal to the voltage across R_4 at the junction of R_3 and R_4. Because the

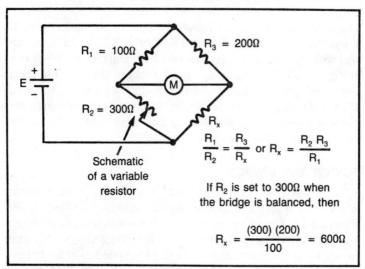

$$\frac{R_1}{R_2} = \frac{R_3}{R_x} \text{ or } R_x = \frac{R_2 R_3}{R_1}$$

If R_2 is set to 300Ω when the bridge is balanced, then

$$R_x = \frac{(300)(200)}{100} = 600\Omega$$

Fig. 3-24. Example used to determine the magnitude of an unknown resistor, R_x, by utilizing the Wheatstone bridge.

71

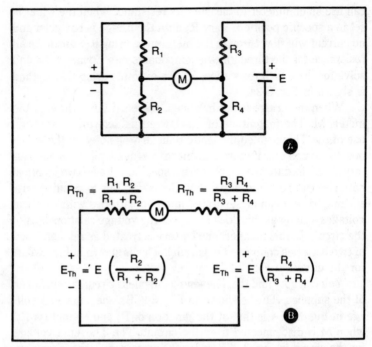

Fig. 3-25. A. Bridge circuit is divided into two parts. Each part has its independent voltage supply. B. Thevenin equivalent circuits used to determine the current in the meter when the bridge is unbalanced.

Thevenin resistance is the resistance seen by the meter when E is shorted, it is R_1 in parallel with R_2 for Fig. 3-25A and R_3 in parallel with R_4 for Fig. 3-25B. The two Thevenin circuits are drawn in Fig. 3-25B. The meter is placed between the two Thevenin equivalent circuits. The resistance of the meter is added to all Thevenin resistances in the circuit. Current flowing through the meter is calculated by dividing the total resistance in the circuit into the total of the Thevenin voltages.

POWER AND ENERGY

In physics, energy has been defined as the ability to do work. Work is not simply a force that you may apply to an object. If the object does not move over a distance, no work is being done. Work is being done or energy is expended if the object moves while you are applying the force. Energy must be supplied in order to perform work.

In physics, energy is changed to a form referred to as work.

Work is a force, F, acting over a distance, s. If there is a large amount of friction between the moving object and the surface over which it is pushed, the force must be large. The required energy must be large in order to overcome the friction and also be able to do the required job. Because of the rubbing of two surfaces due to friction, a portion of the applied energy is used to move the object and the remaining energy turns to heat.

Heat is a form of energy. So are light, electricity, chemicals, and so on. The battery supplies energy because of a chemical reaction in the cells. Steam engines perform because of the energy supplied by heat. Light energy activates photoelectric cells. And electricity is frequently used as a source of energy that is converted into heat, light, and so on. Electric energy is measured in units of *joules*.

We are constantly involved with energy utilized by electric circuits. your bill from the electric company is based upon the amount of electric energy you consume each month. But in circuit work, we deal with power rather than with energy. *Power* is defined as the rate at which work is being done. The amount of energy in joules used each second is power. Power is stated in units called *watts*, abbreviated W.

Just as was the case with volts, ohms and amperes, prefixes are used to indicate power of different magnitudes. Megawatts is the term used for millions of watts so that 1,000,000 watts = 1 megawatt, or 1 MW. Similarly, 1000 watts = 1 kilowatt = 1 kW; 1/100 of a watt = 1 milliwatt = 1 mW, and 1/1,000,000 of a watt = 1 microwatt = 1 μW.

You pay the electric company a specific amount of money for each unit of energy you use. Energy is computed by these companies as the product of watts and hours. This is energy because it is the power consumed multiplied by the time during which the power is being used. The electric companies refer to this more precisely as watt-hours of energy. They may divide this number by 1000 to get the energy in terms of kilowatt-hours. So if you use a 100-watt electric light for 72 hours during the month, you used 72 × 100 = 7200 watt-hours of energy or 7.2 kilowatt-hours of energy. If the price of the electric energy is 20¢ for each kW hour, the bill would be 0.20 × 7.2 = $1.44.

Energy is dissipated in a resistor when current flows through it. The amount of energy dissipated each second is power. A resistor is rated by the amount of power it can dissipate. There are resistors that can dissipate less than 1/4 watt. Other resistors can dissipate hundreds of watts. When selecting a resistor for use in a circuit, it is logical to calculate the power that it must dissipate due to the current flowing through it or to the voltage across it. For safety's sake, the resistor used in the circuit should be capable of dissipating about double the calculated power. Thus, if your calculations indicate that the resistor in the circuit dissipates one-half watt, you should use a one-watt component.

The amount of power being dissipated can be determined through use of Equation 3-4. If you know the voltage, V, across the resistor and the current, I, flowing through it, the power, P, that the resistor dissipates is

$$P = VI \qquad\qquad \textbf{Eq. 3-4}$$

Just as was the case with Ohm's law, the power equation can take three forms. The second version of this equation is

$$P = I \times I \times R = I^2R \qquad\qquad \textbf{Eq. 3-5}$$

because I^2 is exponential notation indicating that I is multiplied by itself. (If I^4 were in the equation, it would indicate I multiplied by itself four times, or $I \times I \times I \times I$.) The third form of the equation is

$$P = \frac{(V \times V)}{R} = \frac{V^2}{R} \qquad\qquad \textbf{Eq. 3-6}$$

Equation 3-4 is used if you know the voltage across the resistor and the current flowing through it. Equation 3-5 is used if you know the current and resistance but may not be aware of the voltage across the resistor. Equation 3-6 requires a knowledge of both the voltage across the resistor as well as the component's resistance.

In Fig. 3-26, three resistors are connected across a 12-volt supply. We do not, offhand, know the quantity of current flowing through R_1, although if you wanted you could calculate it using Ohm's law. Be you can determine the power it dissipates without knowing the current by simply knowing the resistance and voltage. power dissipated by R_1, determined by using Equation 3-6, is $12^2/20 = 144/20 = 7.2$ watts. Power dissipated by R_2, determined by using Equation 3-4, is $12 \times 1 = 12$ watts. Power dissipated by R_3, determined by using Equation 3-5, is $6^2 \times 2 = 72$ watts.

Similar methods can be used when determining the power dis-

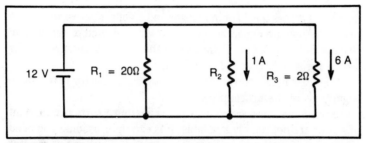

Fig. 3-26. Circuit with three resistors connected in parallel. The power dissipated by each resistor is determined.

sipated by resistors in a series circuit. Several different procedures can be used to determine dissipation by the resistors in the circuit drawn in Fig. 3-27. In one method, we can start by determining the current in the resistors. Because the total circuit resistance is 20 ohms + 4 ohms = 24 ohms, $I = V/R = 12/24 = 1/2$ ampere. Using Equation 3-5, the power dissipated in the 20-ohm resistor is $(1/2)^2(20) = (1/2)(1/2)(20) = (1/4)(20) = 5$ watts, while the power dissipated in the 4-ohm resistor is $(1/2)^2(4) = 1$ watt.

In another procedure you can use the voltage-divider equation to determine the voltage across each of the resistors. The voltage across the 4-ohm resistor is $[4/(20 + 4)]\,12 = 2$ volts. Voltage across the 20-ohm resistor is 12 volts − 2 volts = 10 volts. Using Equation 3-6, the power dissipated by the 20-ohm resistor is $10^2/20 = 5$ watts, and the power dissipated by the 4-ohm resistor is $2^2/4 = 1$ watt.

To use the third procedure you determine the current through the resistors as well as the voltages across the resistors. Using this calculated data, current through both resistors is 1/2 ampere while

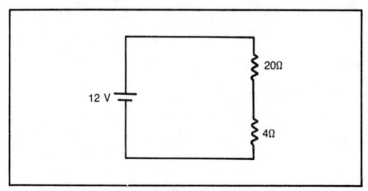

Fig. 3-27. Power dissipated by resistors connected in series.

10 volts is across the 20-ohm resistor and 2 volts is across the 4-ohm resistor. Applying Equation 3-4, the power dissipated by the 20-ohm resistor is $10 \times 1/2 = 5$ watts, and the power dissipated by the 4-ohm resistor is $2 \times 1/2 = 1$ watt.

The calculated power dissipated by each resistor is the same regardless of the procedure used.

Sometimes resistors capable of dissipating a particular amount of power are not readily available. For example, assume you need a 100-ohm, 4-watt carbon-deposited resistor. This type of resistor is made only with dissipation capabilities of up to 2 watts. To fulfill the 4-watt requirement, two 2-watt resistors can be used either in a series or in a parallel arrangement. Because these are both two 2-watt components, each can dissipate 2 watts to satisfy the required capability of a total dissipation of 4 watts.

When wired in series, the two identical resistors should add up to 100 ohms. Two 50-ohm resistors can be used to satisfy this requirement. a circuit will usually behave properly if two 47-ohm resistors are used instead of two 50-ohm resistors to produce a total of 94 ohms rather than 100-ohms. This compromise may be necessary because 50 ohms is not a standard value while 47 ohms is readily available.

If a parallel combination is preferred, two 200-ohm, 2-watt resistors can be so connected to form the desired 100-ohm, 4-watt component. This can be derived using Equation 3-2, which indicates that the equivalent resistance of two 200-ohm resistors connected in parallel is equal to one-half the value of each resistor or (200 ohms) (200 ohms)/(200 ohms + 200 ohms) = 100 ohms.

Chapter 4

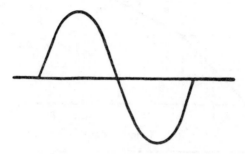

Alternating Current

Each voltage supply described above and shown in the associated drawing, has a specific polarity at its output terminals. One terminal of the supply is always positive and the second terminal is always negative. Current flows in one direction through a circuit when it is powered by this type of supply. Voltage never changes its polarity. When these conditions exist, the source of current and voltage is referred to as a *dc supply*, where dc is an abbreviation for *direct current*.

Supplies that keep reversing in polarity and current flow direction are also available. These sources are designated as *ac supplies*, where ac is an abbreviation for *alternating current*. Although the battery is a dc supply, ac generators and electronic circuits supplying ac are also quite commonly available. The voltage and current supplied by an electric company to its customers is seldom anything other than the ac variety.

GRAPHS AND THINGS

Ac waveforms and characteristics are best understood through use of specific individual graphs as well as through combinations of different types of drawings. A graph is essentially a drawing showing how two items that vary simultaneously are related to each other. In Fig. 4-1, you see how the voltage in a specific circuit can vary with the passage of time. It shows specific voltages at various instances of time.

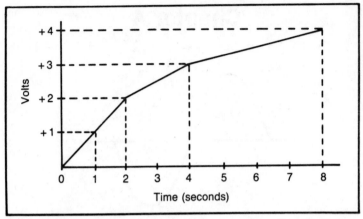

Fig. 4-1. A graph showing how voltage in a circuit may vary with respect to time.

The plot starts at the time when the voltage is at a specific level (0 volt in this graph). The voltage level at specific times after this initial start can be determined from the curve. Time is indicated on the horizontal line or axis. This is referred to as the x axis. Numbers at this axis indicate a quantity of seconds. There is a 0 at the point where the horizontal axis crosses the vertical line or y axis. All events are considered to begin at this time and not before. This point is assigned the number 0 to indicate that it is the time at which we begin our study of the voltage curve. The point marked 1 on the horizontal axis represents the time at one second after the starting point. Similarly, the points numbered 2, 3, 4 and so on indicate the times at 2, 3, and 4 seconds after the start. Because the plot of time is linear, the time between any two listed points can be estimated. Thus, 1 1/2 seconds after the count started is midway between the 1-second and 2-second markings on the scale. The 3.7-second point is between the 3-second and 4-second markings. Just add 7/10 of the distance between the two markings to the 3-second marking to find the point representing 3.7 seconds.

The scale on the vertical axis is in volts. In this drawing, the scale starts at 0 volt and continues to +4 volts. The scale and starting points depend upon the voltage to be displayed on the graph. Thus, if you were not interested in voltage variations below the 2-volt level, but only in voltage variations between 2 and 4 volts, the scale could have started at +2 volts and go up to +4 volts instead of covering the entire range from 0-volt to +4-volts. The same variation could have been used when considering the calibration of the horizontal axis. Here, if only the portion of the curve under study involved the time between 5 seconds and 6 seconds after the voltage was initially applied, the scale for the horizontal axis could have started with the 5-second point rather than with 0 second and ended with the 6-second point. If that 1-second period was spread over the full length of the horizontal axis, points indicating portions (or 1/10) of a second could easily have been plotted between the 5-second and 6-second notations.

Getting back to Fig. 4-1, note that vertical lines have been drawn from several points on the horizontal axis. These lines represent the time indications on the x-axis. Similarly, horizontal lines have been drawn from voltage points on the vertical axis. These represent the voltages noted on the y axis.

The curve shows how the voltage varies with time. It crosses the axis at the 0-second/0-volt point. At the time when the plot begins at 0 second in time, 0 volt is available.

The vertical line from the 1-second designation point on the horizontal axis crosses the voltage curve at a specific point. By drawing a line from that point to the vertical axis, you can see that the point is at the 1-volt level. Thus, at the 1-second instant in time 1 volt is available.

Following a similar procedure, +2 volts is available at 2 seconds in time, +3 volts at 4 seconds in time, and +4 volts at 8 seconds in time. No information is provided as to how this voltage changes after eight seconds has elapsed. If you want to determine the voltage at the 2.5-second instant, draw a vertical line from a point midway between the 2-second and 3-second markings on the horizontal axis and note where that line crosses the voltage plot. Then draw a horizontal line from that point of intersection to the vertical axis. It is about one-fourth the way (or 0.25 of the total distance) between the +2-volt marking and the +3-volt marking. Voltage at 2.5 seconds is about 2.25 volts.

Plot Angles in a Circle

The point at the center of a circle in Fig. 4-2 is labelled O or origin. The *radius* is any line from this center point to the circumference of the circle. (The *circumference* is the actual circle or the curved line forming the circle.) Lines O to C, O to A, and O to B are all radii. (Radii is the plural of the word radius.) All radii are the same length and are equal to exactly one-half the length of the diameter.

The *diameter* is a line that connects two points on the circumference of the circle. But that line must pass through the O center point O. Consequently, it is the longest line that can be drawn in a circle. Here, it is shown as a horizontal line extending from point A to point B. It can be drawn at any angle with respect to the horizontal so long as it passes through the o point and connects two points on the circumference. Regardless of the angles they form with the horizontal, all diameters are the same length and are equal to twice the length of the radii.

The horizontal radius OB shown here is considered as being at zero degree (0°). All other radii form angles with this horizontal radius. The angle formed by any two radii starts at the center point of the circle. Radii OC and reference OB form an angle using this 0 point. This is referred to as angle COB. When radius OC is directly over radius OB, that angle formed with the reference line OB is 0°.

Radius OB is considered as being fixed in place. When another radius is drawn here, such as OC, it is also fixed in place. One setting is shown for radius OC. It is drawn in one spot, but it may be considered as a radius which can rotate inside the circle, continuously changing the angle COB. It may be imagined as rotating in a counter-clockwise direction. It starts to rotate from the 0° point where it is directly over the OB radius. It can continue to rotate in this counter-clockwise direction until it reaches and goes directly over the OA radius. But it does not have to stop there. It may continue rotating indefinitely. After OC has completed one full rotation through the entire circle and returns to its starting point, it is referred to as having completed a *full cycle*.

Suppose rotation starts when OC is directly over OB, and the angle formed by the two radii is 0°. As OC keeps rotating, angle COB increases from the 0° starting point. When it has completed the full cycle to once again reach its starting or 0° point and be over the OB radius, it is said to have rotated through an angle of 360° (360 degrees). Because the OC radius is shown in the drawing at only 1/8 of that angle, in this location it forms an angle of 1/8 of 360° or an angle of 45° with OB. As it keeps rotating, angle COB increases until one cycle of rotation has been completed through 360°. If it rotates through a second full cycle, it has gone through an angle of 360° × 2 = 720°. Similarly, if it went through sixty of these cycles, it has rotated through 360° × 60 = 21, 600°. Should the time for 60 cycles be accomplished within one second, you can say that the radius rotated through 60 cycles per second or 21, 600°/second. (The symbol / indicates "per.")

When working with numbers, very few, if any, categories in-

Fig. 4-2. Angles in a circle formed by its radii.

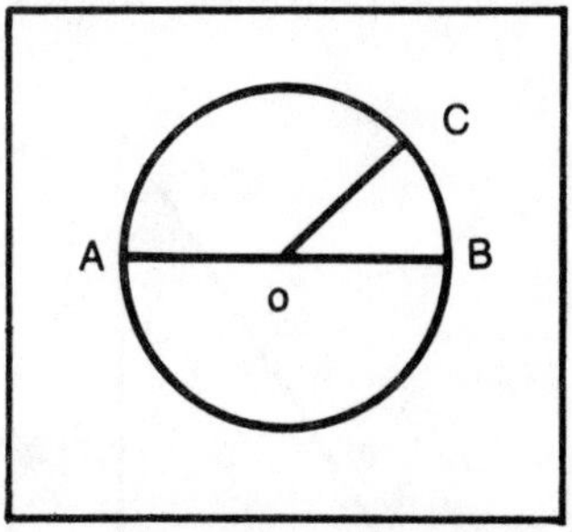

volve one system of units. When measuring length, you use either the "mile" or the "meter" as the unit. When you measure weight, it is either in units of "pounds" or units of "kilograms." Even money may be considered in "dollars" or in "yen." About the only factor that holds in all systems is time, measured in units of seconds. So as you could expect, angles (the symbol is∢) are not measured only in degrees. It has been found very useful to refer to angles in terms of a number of radians.

In Fig. 4-2, when the OC radius travelled through 360°, you could just as easily have said that it rotated through 2π radians. The Greek letter π (pi) represents the number 3.1416. Usually 3.14 is accurate enough for most purposes, dropping the last two digits.

When the OC radius was originally over the OB radius, the angle was noted as being 0°. This is the same as saying 0 radians. At the 45° angle shown in the drawing, the OC radius travelled 1/8 of its full cycle, so that angle if forms with OB is $1/8 \times 2\pi = \pi/4$ radians. To convert from an angle in degrees to an angle in radians, multiply that angle by 2π and divide it by 360. Thus, 90° is $(2\pi)(90)/360 = \pi/2$ radians. Working backwards, if you know the radians and wish to convert the notation to degrees, simply multiply the number in radians by 360 and divide it by 2π. So if you start with $\pi/4$ radians, its angle in degrees is $(\pi/4)(360)/2\pi = 45°$. But these conversion formulas are not necessary if you simply remember that 360° = 2π radians and take it logically from there.

Sine of an Angle

The relative lengths of the sides of a triangle (or figure containing three angles and three sides) are related to its angles. If you consider the triangle in Fig. 4-3, the angle at the lower right corner is 90°. Because one angle is 90°, the figure has been given the special name of *right triangle*. The side opposite the right or 90° angle has been labelled h and referred to as the *hypotenuse*.

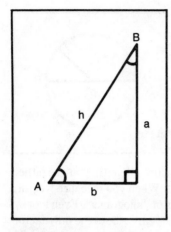

Fig. 4-3. A right triangle.

The other two sides of the triangle, a and b, are opposite angles A and B. In this discussion, I consider a, b and h as the lengths of the sides of the right triangle and A and B as the number of degrees of the two other angles.

The ratio of the lengths a to h is referred to as the *sine* of the angle opposite side a. It is written mathematically as

$$\sin A = \frac{a}{h} \qquad \text{Eq. 4-1}$$

For angle B, its sine is obviously b/h because side b is opposite angle B.

Another relationship you will be using in the discussion of electricity and electronics is the fraction formed by the length of a line adjacent to the angle with the length of the hypotenuse. The term for this is *cosine*. Stated mathematically,

$$\cos A = \frac{b}{h} \qquad \text{Eq. 4-2}$$

so that cos B = a/h. Notice how cos B and sin A are identical, while cos A is also equal to sin B.

In addition, there is also a term for relating the length a to the length b, while ignoring h. It is referred to as the *tangent* of an angle. For angle A, it is written as

$$\tan A = \frac{a}{b} \qquad \text{Eq. 4-3}$$

Obviously, tan B = b/a. If you do a little simple algebra, you will see that sin A/cos A = tan A because (a/h) / (b/h) = a/b. Similarly, sin B/cos B = tan B.

Another equation useful in electronics work is the relationship between the lengths of all sides of a right triangle. It is

$$a^2 + b^2 = h^2 \qquad \textbf{Eq. 4-4}$$

showing that the sum of the squares of the lengths of two of the sides of the right triangle is equal to the square of the length of the hypotenuse.

Fractions a/h, b/h, and so on, are merely the ratios of the length of one side of a right triangle to the length of a second side. The sine, cosine, and tangent of the angle are numbers indicating these ratios. Tables are available in math books that list the ratios for different angles. For example, you will find from the table that the sine of a 60° angle is 0.866. So if A is 60° or $\pi/3$ radians, the ratio of length a to length h is 0.866. Consequently, sin A = 0.866 and a/h = 0.866. Should h be 10″ long, the length of a would be 10(0.866) = 8.66″.

Sine Waves

You can now relate the definition of the sine of an angle to lengths of lines drawn in a circle. Consider the drawing in Fig. 4-4, which is simply the circle shown originally in Fig. 4-2. Here, more radii are shown. Each radius touches the circumference of the circle. A vertical line is drawn from the point where the radius touches the circumference to the horizontal line AB.

Consider each radius as the hypotenuse of a right triangle. Each triangle is formed by three lines. These lines are 1) the radius in question; 2) the vertical line that extends from the point where the

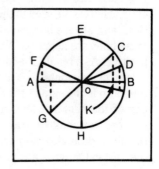

Fig. 4-4. Circle showing various radii.

radius touches the circumference down to the horizontal line AB; and 3) the length of the portion of the horizontal line that is between the o point and the point where the vertical line touches it. One of the triangles in Fig. 4-4 that is formed by these three lines involves radius OC. Here the lines are 1) radius OC; 2) vertical line CK; and 3) line OK. Despite the fact that all radii in a circle are of equal lengths, the lengths of the vertical and horizontal lines of the triangle depend upon the angle formed by the radius under consideration with the horizontal line.

The sine of the angle formed by a specific radius with line AB is equal to the length of the vertical line divided by the length of the radius. For convenience, let us consider the lengths of the radii as being equal to unity or 1.

Starting with the radius OD in Fig. 4-4, the length of the vertical line from point D to the line OB, is the sine of angle DOB. This is because the sine of that angle is equal to the length of the line from D to line OB divided by the length of the radius. Since the radius is 1, the length of that vertical line is simply equal to the sine of the angle.

The sine of the larger angle, COB, is the length of the vertical line between C and OB. This length is larger than the one from D to the horizontal line, so the sine of this angle must be greater than the sine of the angle in the previous case.

When the angle is EOB or 90°, the length of the vertical line is at a maximum and equal to 1 (1 is the length of the radius). As the angle increases above 90°, the sine of the angle decreases. For example, the line from F to the horizontal line is the sine of angle FOB. The sine of angle FOB is identical with the sine of the smaller angle, FOA. The sine of the angle reaches zero when the angle is 180° or π radians. At this angle, the radius under consideration is coincident with or lies over radius OA.

The sines of all angles are negative when the angles exceed 180°. This can be seen in the figure. Here, the vertical lengths from G, H, and I to the horizontal are all below line AB. Because of this, these lines are negative. The radius is always considered as being positive. As a result, the ratio of a negative vertical line in the area below the horizontal axis to the positive radius is negative. Because this ratio is the sine of the angle, the sines of angles formed using any of the negative lines must also be negative.

You can make a drawing to indicate how the magnitudes of the sines of angles vary with the degrees of the angles. Because the vertical lengths are proportional to the sines of the angles, a drawing showing the relative lengths of the vertical lines at the different angles is identical with a drawing showing how the magnitudes of the sines of the angles vary with the degrees of the angles. This is shown in Fig. 4-5. The height at each angle is proportional to the vertical lengths shown in Fig. 4-4 for that angle. This drawing is referred to as one cycle of a sine wave because it contains information about the sines of angles between 0° to 360°, or between 0 radians to 2π radians.

Looking at the sine wave, several factors should be noted. All points correspond to the different vertical amplitudes at the various angles. Instead of relating the amplitudes to angles, they can be related to time. Let us say it takes one second for the radius to rotate from 0° to 360° to complete one cycle of the sine wave. So if it starts at time 0 at 0°, its amplitude or height is at a positive peak at the instant that one-quarter/second of time has elapsed. The amplitude drops to 0 at 180° or at the one-half/second mark and decreases to a negative peak after another one-quarter/second has passed. It is at the negative peak at the three-quarter/second point or at 270°. The cycle has been competed when the amplitude rises from the negative peak to a zero level. This is at an angle of 360° or at the moment that 1 second in time has passed. This same cycle is repeated for the next second between 360° and 720°, for the third second between 720° and 1080°, and so on.

A cycle does not have to take only one second. It can occupy much more time than one second. If it occupies five seconds, each

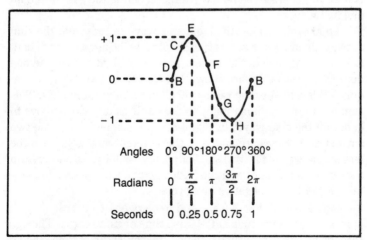

Fig. 4-5. Sine wave using data determined from Fig. 4-4. All letters correspond to points labelled in Fig. 4-4.

length of time would be five times that shown here, so that at 180° or π radians, the time would be 2.5 seconds.

However, in electronics each cycle usually occupies much less than one second. You can have ten or even billions of these cycles in one second. Other items worthy of note are the notations used to indicate the positive and negative peak amplitudes of the sine wave. It is shown to vary from a maximum of $+1$ to a minimum of -1. The numbers "1" were chosen for convenience and are identical with the numbers chosen to indicate lengths of the associated radii in the circle in Fig. 4-4. Magnitudes of the sine wave at all angles are related to these positive and negative maximums.

Amplitudes at all points on the curve are identical with the vertical heights from the tip of the radius to the horizontal diameter at specific angles in the circle. These heights vary as shown on the curve. The amplitudes are one-half of the positive or negative peaks when they are at 30°, 150°, 210° and 330°.

The cosine waveshape is identical to the sine wave except that it starts at the peak point, E, rather than at B. Point E is considered as being at 0°; point F is at 90°, and so on.

If two sine waves with different amplitudes are available at the same time, they can be added to form one curve that is equivalent in magnitude to the total of the two sine waves. This can be done by considering the lengths of the radii in the two circles used to generate the individual sine waves, and then finding the sum of these radii. The radius derived from the sum of the original radii is used to generate the new wave representing the sum of the individual waves.

Another method of finding a sine wave to represent the sum of two individual sine waves is to simply add the amplitudes of both sine waves at each of the angles. This method, as well as the one discussed previously, are shown in Fig. 4-6. In Fig. 4-6A, the radius OB is added to the radius OC to get a total radius OD. This sum of the two radii is rotated in the circle as described above to generate the sine wave. In Fig. 4-6B, the magnitudes of the two smaller sine waves are added at the different angles to form the larger sine wave. The two smaller positive half-cycles add to form a larger positive half-cycle and both negative half-cycles add to establish the larger negative half-cycle.

Sine waves in Fig. 4-6 are said to be *in phase* because they all start at a zero amplitude at the same time or same angle. They go through an entire cycle in the same *period* or length of time. Should the two cycles be out of phase, the waves would be at the zero level

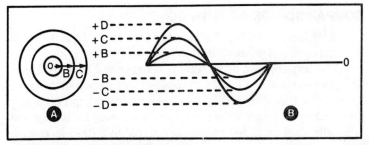

Fig. 4-6. Sum of two in-phase sine waves.

at different angles. For example, sine wave OC may start 45° before sine wave OB. The two curves may then look like those drawn in Fig. 4-7B. This phenomenon is described as sine wave OC *leading* sine wave OB by 45°.

In Fig. 4-7B, the amplitudes were added at each angle as in Fig. 4-6B. In Fig. 4-7A, the amplitudes were added vectorally using the radii as in Fig. 4-6A. Here, however, radius OC is drawn as in the previous figure, but OB is shown as being at an angle of −45° (or +315°) with respect to the horizontal. This is because the OB curve reached the 0° point 45° after the OC curve, thereby lagging behind by 45°. A parallelogram is formed using lengths OB and OC to determine the magnitude of the sum, OD, as well as the angle this line representing the sum forms with the 0° horizontal line. The line, OD, is the sum of the two out-of-phase sine waves. The angle, COD, is the angle by which the total or sum of the two out-of-phase curves lags behind the curve formed due to the OC radius. The total curve and the lagging angle are also shown in Fig. 4-7B.

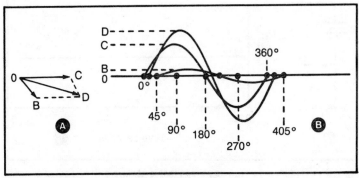

Fig. 4-7. Two out-of-phase sine waves are added. The sum and phase of the total of the two waves are derived from A in this figure.

COMPARISON OF AC WITH DC

Using this knowledge of graphs and sine waves, I can now turn to a discussion of what ac is all about.

A battery supplies dc voltage. It is shown in Fig. 4-8A. Here, the upper terminal of the battery is positive with respect to the lower terminal. If the lower terminal is connected to ground or is at a 0-volt reference potential (potential is another word for voltage and is usually used when the voltage is referred to a fixed level such as ground) the upper terminal is E volts above the 0-volt level. This is shown graphically in Fig. 4-8B.

Should the battery be reversed in the circuit so that the plus terminal is at ground, the voltage at the minus terminal is E volts below the 0-volt level. This is shown graphically in Fig. 4-9. The voltage remains fixed at $-E$ volts for the full time during which the battery is connected to the circuit with the polarity shown on the graph.

Taking this one step further, assume the circuit in Fig. 4-10A exists. Terminals marked 1, 2, and 3 constitute a switch. When it is set in one position, terminal 1 is connected to terminal 3. In the alternate setting, terminal 2 is connected to terminal 3. Voltage at the output of this circuit is between terminal 3 of the switch and ground.

When terminals 1 and 3 are connected to each other, the out-

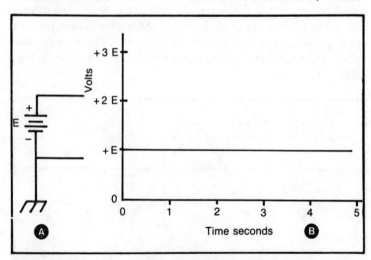

Fig. 4-8. A. Battery supplying dc voltage that is positive with respect to ground. B. Graph showing how this voltage varies with time. In this case it is constant and above the 0-volt axis because it is positive.

88

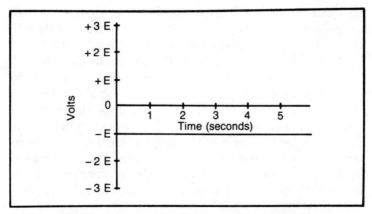

Fig. 4-9. Voltage is negative and at E volts with respect to ground.

put is as shown in Fig. 4-8B. Voltage at one output terminal of this circuit is + E volts with respect to ground. In the alternate setting of the switch, the output is – E volts with respect to ground. This is shown in Fig. 4-9.

Now assume you keep on switching back and forth between one position and the other position of the switch. Then the voltage at the output of the circuit would keep shifting back and forth from + E to – E and then back to + E. This continues as long as the switch is thrown from one setting to the other. If this is done once each second, voltage at the output is as shown in Fig. 4-10B. Voltage is no longer fixed at one polarity, positive or negative, but it keeps on alternating from one polarity to the other. This is the basis of an ac supply. Not only does the voltage keep alternating in polarity, but the current also keeps changing its direction of flow. The voltage is referred to as an *alternating voltage* and the current is an *alternating current*, abbreviated ac. The voltage is most frequently referred to as *ac voltage*. It should be noted that however correct the term alternating voltage is and however technically incorrect the term ac voltage is, the latter term is most commonly used, and I will use it here. Direct current and *dc voltage* are terms applied in a similar fashion when referring to current and voltage that are maintained at one polarity. Voltage and current with this characteristic is, of course, available from a single battery.

Ac Waveform

When the voltages are switched as shown in Fig. 4-10, you end up with a rectangular or *square waveform*. Voltage remains fixed at a positive or negative peak level. It is assumed here that the voltage jumps instantly from maximum level to the other. In real life this is never the case, although the situation may be approximated.

Voltage supplied by the electric companies is not in the form of a square wave. It has the sinusoidal shape shown in Fig. 4-5. Voltage rises gradually to a positive peak, drops slowly through a zero level to a negative peak, and then once again rises slowly to repeat the cycle. If the voltage supply is connected across a resistor, the current flowing through the resistor assumes this identical shape. Because the current flowing in the resistor rises and falls in time with the applied voltage, the current is said to be *in phase* with the voltage.

If the meter shown in Fig. 2-5 were considered in another fashion, it could conceivably be used as a device to generate a sinusoidal ac voltage. Start by assuming that the spring and pointer do not exist. The two leads from the coil do not hang free. They are oriented so that they make connections with contacts mounted on a surface, as shown in Fig. 4-11. The armature is rotated by some force. As it rotates, the leads keep making connections with the fixed contacts.

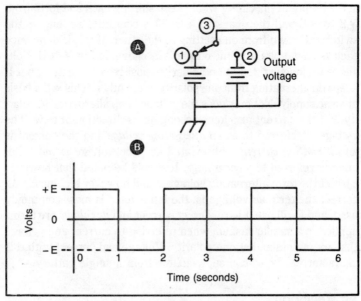

Fig. 4-10. Output voltage polarity altered by use of a switch and two batteries. A. Circuit. B. Output voltage.

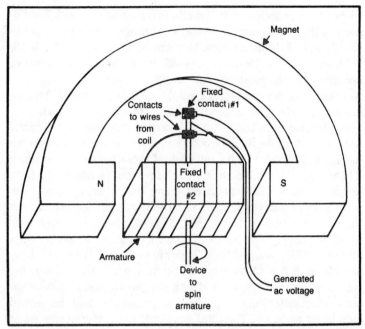

Fig. 4-11. Basic ac voltage generator. Fixed contacts rub on contacts connected to wires and mounted on rotor. The armature is rotated by some mechanical device.

As the coil rotates, it keeps breaking the magnetic field formed by the permanent magnet. A current is induced in the coil. Consequently, a voltage is available across the leads from the coil and at the fixed contacts. The instantaneous magnitudes of the voltage and current depend upon the number of magnetic lines in the field that are broken by the rotating coil. Because the number of lines broken depends upon the instantaneous orientation of the coil, the magnitude of the voltage at the fixed terminals varies with time. These variations take a sinusoidal waveform.

Average and Effective Voltage and Current

When you take an average of a group of numbers, you usually add the numbers and divide the result by the quantity of numbers involved. Thus, if you want to know the average of these six numbers, 4, 27, 10, 16, 13, and 2, you must add all numbers for a total of 72 and divide the sum by 6 for an average value of 12. This average is known as the *mean*.

All numbers in the example shown are positive. But what if

some numbers are positive and others are negative? Let's find the mean of the above group of numbers by changing the signs (or polarity) of some of these numbers. Use numbers -4, -27, $+10$, $+16$, $+13$, and -2. The total is $-33 + 39 = +6$. Because six numbers are involved, the mean average is $+6/6 = +1$.

Now consider the average amplitude of a sine wave. Assume it has the shape of an applied voltage with a peak amplitude of E volts. To determine this average, you must consider one full cycle. You can start by dividing the 360° into twelve equal parts, as shown in Fig. 4-12. Draw a line from each of these points to the curve. There would be twelve lines, but no lines are shown at the 180° and 360° points. This is because the amplitudes here are zero. Now add the amplitudes or heights of all the lines hitting the curve in the section between 0° and 180°. This sum is positive. Continue by adding the amplitudes hitting the curve in the section between 180° and 360°. Since all these lines are negative, the sum of these amplitudes is a negative number. Because both halves of the sine wave are identical, the sums of both groups of amplitudes are identical. They differ only in one sum being positive and the second sum being negative. The total amplitude of an overall sine wave is zero because the sum of a positive number with an identical negative number, is zero. Dividing this by 12 because 12 amplitudes are considered, leaves you with the information that the average amplitude of an entire sinusoidal cycle is equal to zero. So whether the cycle represents current or voltage, the average current or voltage present over a complete cycle is zero.

Note that the sine wave was divided into twelve parts in Fig. 4-11. It could just as easily have been divided into more sections.

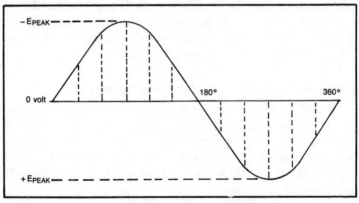

Fig. 4-12. Method of determining mean or average level of a sine wave.

The average number for each positive or negative peak achieved by considering more and consequently smaller, parts is more accurate. Here, less parts were considered because of convenience and ease of explanation. In any event, the *mean* average voltage or current of an entire sinusoidal cycle, is zero.

If we consider the case when a voltage or current is present during only one half of a sinusoidal cycle, an actual average does exist that differs from zero. Here, the amplitudes in the first half of the cycle are noted as they were above. All amplitudes in the second half of the cycle are zero. To determine the mean amplitude for the entire cycle, just add the six amplitudes that exist in the active half-cycle and divide that sum by twelve. The sum must be divided by twelve because there are six amplitudes of zero magnitude in the second half-cycle that cannot be ignored. They must be added to the six amplitudes in the first half-cycle. If the curve discussed here were divided into an infinite number of sections to enable you to get an exact average, you would find it to be

$$e_{AV} = \frac{E_{Peak}}{\pi} \qquad \textbf{Eq. 4-5}$$

where E_{Peak} is the peak amplitude. I will not show how this infinite division and averaging is done, but it is possible with calculus. E_{Peak} can be positive or negative. Its polarity depends upon the half-cycle being considered as active. This mean type of average is applied when rectified power supplies are being considered.

Another type of average considers the sum of the squares of all amplitudes. Unlike the mean average, this average does not ignore the fact that the total power dissipated in a circuit is due to voltage and current present in both halves of the cycle. Twelve amplitudes are involved when determining a close approximation to this average. Six are positive and six are negative. Unlike the procedure used to find the mean of the 360° curve, each amplitude is squared (or multiplied by itself) before being averaged. Thus, if an amplitude is +3 volts, the square of 3 volts is $3^2 = 3 \times 3 = +9$. In the half of the sine wave between 180° and 360° the amplitudes are negative. But the square of each amplitude is positive for it is the product of two negative numbers. Thus, if the amplitude here is −3 volts, $(-3)^2 = (-3) \times (-3) = +9$. All the squares of the individual amplitudes are added and the sum is divided by twelve to determine the average value of these squared numbers. The square root of this average is then determined to establish what is referred to as the *root mean square* average of the amplitudes. The abbreviation for this is rms. If there were an infinite number of sections of the sine wave used to determine the rms average, using calculus you would get an exact rms number equal to

$$E_{RMS} = \frac{E_{Peak}}{\sqrt{2}} = \frac{E_{Peak}}{1.414} = 0.707\ E_{Peak} \qquad \textbf{Eq. 4-6}$$

Root mean square voltage and current are important when analyzing ac circuits. If a dc voltage produced a specific amount of heat in a resistor, the rms ac voltage must be numerically equal to the dc voltage if it is to do the same job. Thus, if 10 volts of dc brings the temperature of a resistor up to 55° Celsius, it would take 10 rms volts of ac across the same resistor to bring its temperature up to the same 55 °C. The peak ac voltage must then be $10 \times \sqrt{2}$ = 14.14 volts if it is to supply the 10 rms volts.

Similarly, dc and ac power and current correspond to each other if the rms values are used for each. As for current, the rms values are:

$$I_{RMS} = \frac{I_{Peak}}{\sqrt{2}} = 0.707 I_{Peak}$$

Rms power is derived by multiplying the rms voltage by the rms current.

$$P_{RMS} = (0.707 E_{Peak})\ (0.707 I_{Peak}) = 0.5\ V_{Peak} I_{Peak}.$$

Frequency

The duration of the single cycle shown in Fig. 4-5 is from 0 to 2π radians. The time for a cycle differs with the application in which the ac is being used. If one cycle is completed in one second, the sine wave is referred to as having a frequency of one cycle per second or 1 cycle/second. In modern notation, the term cycle/sec has been assigned the name *Hertz*, abbreviated Hz. Thus, if there is 1 cycle/sec, it is the same as saying that the frequency is 1 Hz. If there are ten of these cycles during each second, the sine waves are referred to as having a frequency of 10 Hz.

Ac power is supplied by electric companies. In some places in the world, the frequency of the power can be 25 Hz or 50 Hz. Over most of the United States, the power line frequency is 60 Hz. This means that the electric company is supplying 60 of these cycles each second, 60 Hz × 60 secs or 3600 cycles each minute, and 3600 cycles × 60 = 216,000 cycles each hour. That sounds like a lot of cycles, but in reality 60 Hz is a very low frequency.

When you listen to sound from the radio, hi-fi, or television set, you hear these cyclic variations in various forms. They comprise music, speech, sound effects, and so on. This audible material can be composed of frequencies between 16 Hz and 15,000 Hz. It often ranges to above 20,000 Hz. As we grow older, the frequencies that the human ear can hear, fall far below 20,000 Hz, possibly to as low as 6,000 Hz. But the higher frequencies are there regardless of our capabilities of hearing them. In the hi-fi world, a range of 20 Hz to 20,000 Hz is considered the minimum necessary frequency response.

The highest audible frequency, 20,000 Hz, is in reality a very low frequency. Consider radio frequencies. Frequencies from about 550,000 Hz to 1,600,000 Hz are transmitted on the AM broadcast band. Citizen's Band radio operates at about 27,000,000 Hz. FM broadcasts are transmitted on frequencies between 88,000,000 Hz and 108,000,000 Hz; television stations broadcast signals between 48,000,000 Hz and 890,000,000 Hz while radar and other similar devices transmit at frequencies which exceed the 1,000,000,000 Hz range by large factors. The frequency or the number of sinusoidal cycles present each second can be quite high.

Just as in the case with voltage, resistance, and so on, prefixes can be used as abbreviations for large frequency numbers. A 1000 Hz tone is at a 1 kilohertz or 1 kHz frequency. Similarly, 800 Hz and 0.8 kHz are identical frequencies. One million Hertz and 1 megahertz or 1 MHz are identical. A radio station that broadcasts at a frequency of 1200 kHz is supplying 1.2 MHz. The prefix for one billion Hertz is giga so that 16,000,000,000 Hz = 16,000,000 kHz = 16,000 MHz = 16 gigahertz or 16 GHz. If you want to go further, 1000 GHz = 1 THz where THz is an abbreviation for terahertz. Tera is a prefix indicating trillion.

Period

The time it takes for the sine wave to complete the entire cycle from 0 to 2π is referred to as the *period*. Thus, if the frequency is 1 Hz, the time duration of one cycle is 1 second. Should the frequency be double this, or 2 Hz, it takes half of the previously noted time for the sine wave to go through a complete cycle. There are two cycles in each second. Thus, the time for each cycle of a 2 Hz sine wave is one-half second. If you continue with this, you will find that the time for a single cycle at any frequency can be determined using the simple relationship

$$t = \frac{1}{f} \qquad \text{Eq. 4-7}$$

where t is the period in seconds, and
f is the frequency in Hz or cycles/sec.

Thus if a radio station broadcasts at 1 MHz, the period or time duration of each cycle is t = 1/1 MHz = 1/1,000,000 Hz = 10^{-6} secs.

As always, prefixes and abbreviations can be used here. A chart showing a relatively complete listing of these prefixes and abbreviations is shown in Table 4-1. Using this table, the period of a 1 MHz signal is 10^{-6} secs = 1 microsecond = 1 μsec. From this it can be seen that if frequency is considered in kHz rather than in Hz, the period using equation 4-7, is in msecs; if f is in MHz, t is in μsecs; if f is in GHz, t is in nsecs; and if f is in THz, t is in psecs.

TRANSFORMERS

Resistors are passive components that can be used in both ac or in dc circuits. Other components perform their functions only when a variable voltage (or continuously varying voltage such as ac) or current is applied. The transformer falls into this category.

Table 4-1. Abbreviations of Prefixes for Time and Frequency.

| Prefix | Abbreviation | Portion of Seconds or Hertz | |
		Fraction or Multiple	Exponential Form
pico	p	1/1,000,000,000,000	10^{-12}
nano	n	1/1,000,000,000	10^{-9}
micro	μ	1/1,000,000	10^{-6}
milli	m	1/1000	10^{-3}
kilo	k	1000	10^{3}
mega	M	1,000,000	10^{6}
giga	G	1,000,000,000	10^{9}
terra	T	1,000,000,000,000	10^{12}

Structure

Transformers are frequently made with two closely spaced windings. Each winding consists of a coil of wire with a specific number of turns. A voltage is applied to the two ends of one of the windings so that current flows in its turns of wire. As in the case of the coil in a meter movement, a magnetic field is established in the area around the coil when current flows through it. If dc flows through the coil, a magnetic north pole is established at one end of the winding and a magnetic south pole is at the second end. Which end is north and which end is south, depends upon the direction in which the current is flowing.

Should ac be applied to the coil, the magnetic poles at the two ends of the coil keep changing. This is because the direction in which the current flows changes with each half-cycle. This constant reversal of the magnetic field is a basic factor in how a transformer works.

I discussed earlier that a current is induced in a wire if that wire is moved through a magnetic field, breaking that field. Because current flows in this wire, a voltage is developed across it. The polarity of this voltage depends upon the direction in which the current is flowing. It is important to realize that there is no current induced in the wire or coil of wire, if it does not move in the magnetic field, to break it to some degree.

Now let us suppose that a coil of wire is fixed in a magnetic field. It does not move. If the magnetic field is fixed in one direction and at one level, no current is induced into the coil. But assume that we have a field that keeps changing in magnitude and polarity. Instead of the wire moving and breaking the field, the field surrounding the coil keeps shifting. This is identical to the situation noted earlier, but now the magnetic field moves over the wire rather than the wire moving across the field. Because there is a break in the field when it shifts over the coil of wire, a current is induced into the coil. Voltage is developed across the winding because of the presence of the induced current.

A transformer can be constructed of any number of coils. Schematics of different types of transformers with two windings are shown in Fig. 4-13. In all transformers, ac voltage is applied to one winding to establish the varying magnetic field. Because of this varying field, an ac current is induced into the second winding. In Fig. 4-13A, the magnetic field is built up in an iron core. This core is not solid, but is made of strips of metal known as *laminations*. Both coils are wound around one core so that there is good

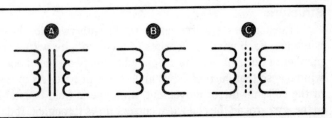

Fig. 4-13. Schematic symbols for transformers. Each transformer shown here has two windings but the cores differ. A. Laminated iron core. B. Air core. C. Powdered iron core.

coupling between the windings. Transformers using this configuration are good at low and audio frequencies, and are usually used at frequencies below 100 kHz. The air core transformer in Fig. 4-13B does not use any magnetic material that is common to both coils except air. Despite the relatively weak magnetic field resulting in poor coupling between windings, this arrangement is necessary in high-frequency applications. A transformer using a powdered iron core is shown in Fig. 4-13C. This type of transformer is also used at high frequencies. Because of the metal, coupling between the two windings is better here than it was in the air core unit. Different types of metals and powders are used where they can best accommodate the frequencies at which the transformers are being applied.

A commonly used structure of a laminated iron-core transformer is shown in Fig. 4-14. Only six laminations or metal strips are shown, although actual transformers usually have many more of these. Both coils are wound around the center of the core.

Assume an ac voltage is applied to winding #1. Because the input voltage is applied to this winding, it is referred to as the *primary*. Winding #2 is the only *secondary* winding in this transformer. Different transformers have several secondaries.

Voltage Ratio vs Turns Ratio

The ac voltage, V_1, across, and ac current, I_1, flowing in the primary winding establish a varying magnetic field in the core. Current is induced into the secondary winding because of this changing field. The amount of voltage, V_2, across the secondary because of the current, I_2, flowing in it is related to the ratio of the number of turns, N_1, in the primary to the number of turns, N_2, in the secondary. It can be stated simply as equation 4-8.

$$\frac{V_1}{V_2} \;=\; \frac{N_1}{N_2} \qquad\qquad \textbf{Eq. 4-8}$$

Suppose you have a transformer with 75 turns in the primary and 300 turns in the secondary. Then $N_1 = 75$ and $N_2 = 300$. If V_1, the ac voltage applied to the primary, is equal to 100 volts, the voltage across the secondary winding, V_2, can be determined through use of Equation 4-8.

$$\frac{100}{V_2} \;=\; \frac{75}{300}$$

Solving for V_2, it is equal to $100(300/75) = 400$ volts.

Current Ratio vs Turns Ratio

The relationship of the currents in the two windings is just op-

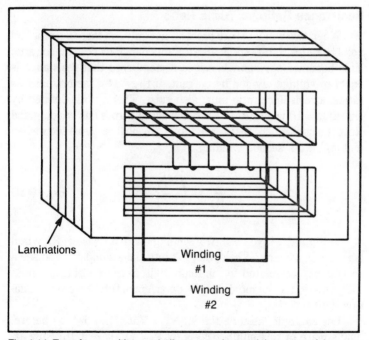

Fig. 4-14. Transformer with two windings wound around the center of the core. Winding #1 is shown using heavier wire than winding #2, but this is not always the case.

posite to that of the voltages. It can be determined using Equation 4-9.

$$\frac{I_1}{I_2} = \frac{N_2}{N_1} \qquad \text{Eq. 4-9}$$

From this equation, it is obvious that less ac current flows in the winding with the greater number of turns than in the one with the lesser turns. If, for example, the current is 4 amperes in winding #1 of the transformer just described, then the current in winding #2, can be determined using Equation 4-9 as follows.

$$\frac{4}{I_2} = \frac{300}{75}$$

Then $I_2 = 4(75/300) = 1$ ampere.

Resistance Ratio vs Turns Ratio

When two windings are present, not only are there current and voltage ratios, but there is also an impedance or resistance ratio. This becomes obvious when it is remembered that resistance is equal to voltage divided by current. If these two factors exist as ratios, a resistance ratio must also exist. This ratio is defined by Equation 4-10 where R_1 is a resistor wired across one winding and R_2 is the consequent resistance (not a physical resistor) seen when looking into a second winding.

$$\frac{R_1}{R_2} = \left(\frac{N_1}{N_2}\right)^2 \qquad \text{Eq. 4-10}$$

In Fig. 4-15A, you see two windings with a resistor, R_1, wired across the primary. When you look into the secondary, you see a resistance, R_2, related to the magnitude of the resistance across the primary. Its magnitude can be determined through use of Equation 4-10.

For example, assume the $N_1/N_2 = 75/300 = 1/4$, as before. The square of this number is $(N_1/N_2)^2 = (1/4)^2 = 1/16$. Now assume that R_1 in the schematic, is equal to 25 ohms. Then $25/R_2 = 1/16$ so that R_2 must be equal to 400 ohms. Thus, the impedance seen by looking into the secondary is greater than that

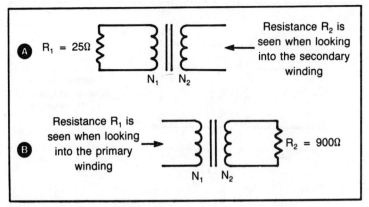

Fig. 4-15. Reflected resistance.

across the primary. This is because in this transformer there are more turns used to form the secondary winding than are used to form the primary. This factor is magnified because the impedances are related not only by the turns ratio, but by the turns ratio squared.

To do one more example, assume 900 ohms is across the secondary winding as in Fig. 4-15B. The primary is open and you want to know what impedance is seen at the primary because of this 900 ohms. Because the same transformer is being used, the turns ratio squared remains at 1/16. The relationship of resistance to turns ratio is $R_1/900 = 1/16$, so that R_1 seen in the primary is 56.25 ohms. As in Fig. 4-15A, the turns ratio squared relationship still holds, but now the reflected resistance is less than that across the secondary. This is because the physical resistor is across the winding with the greater number of turns and is observed as a resistance across the winding with the smaller number of turns.

But this is only part of the story. Because each winding is made up of turns of wire, each turn by itself has a built-in resistance. The resistance of the wire adds to the resistance seen when looking into a winding. Consider Fig. 4-16A. If N_1 should have a winding resistance of 45 ohms, the total resistance across the winding is 25 ohms + 45 ohms = 70 ohms. When observed by looking into the secondary, this resistance appears as a reflected resistance of (70 ohms)(16/1) = 1120 ohms. This is because the turns ratio squared of the primary to the secondary is 1/16. If the secondary winding has a resistance of 20 ohms, the total resistance observed when looking into the secondary winding is 1120 ohms + 20 ohms = 1140 ohms.

Analyzing this same situation with 900 ohms across the secon-

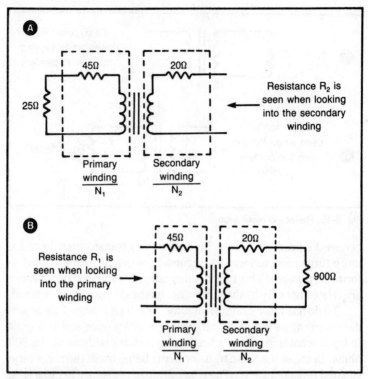

Fig. 4-16. Reflected resistance includes the effects of the resistance of the two windings.

dary winding, the total resistance in that winding is 900 ohms + 20 ohms = 920 ohms. When looking into the primary, the resistance reflected into that winding from the secondary is (920) (1/16) = 57.5 ohms. Adding this to the 45-ohms resistance of the primary winding, 102.5 ohms is seen by looking into the primary.

Power in the Two Windings

The final very important factor to be considered is the amount of power transferred from one winding to the other. If the transformer is 100% efficient, there is no power lost in the windings and core. In the example used in the discussion of current and voltage ratios, it was determined that V_1 = 100 volts, V_2 = 400 volts, I_1 = 4 amperes and I_2 = 1 ampere. Again using this same transformer, you should note that the power in the primary winding, the product of the voltage across it with the current through the winding, is 100 volts × 4 amperes = 400 watts. This is equal to the power in the secondary winding, which is 400 volts × 1 ampere = 400 watts. Power is identical in both windings if the transformer is perfect. But perfect transformers don't exist in the real world, so there are losses.

Efficiency

If all flux linked both windings of the transformer, efficiency would be high and power losses would be at a minimum. But this never happens. There is always some sort of *leakage* where a percentage of the magnetic flux does not remain in the core. Instead, the flux flows out into the surrounding air. This flux never reaches the secondary winding.

Leakage is only one of the factors which involves the core and results in a loss of power between the primary and secondary windings. There would be good efficiency if the magnetic field in the core and the ac voltage applied to the primary winding shifted polarity at the same time. But the metal core tends to hold the initial magnetic polarity. It shifts only after some "urging" from the applied energy. Because of this delayed change, not all power from the primary is induced into the secondary. This is because a portion of the applied power or energy is used to force this shift in polarity, this energy never reaches the secondary and is lost. This is known as a *hysteresis loss*. Different materials are frequently used in the core to reduce this loss. Less energy is required to change the polarity in these materials than is required in pure iron.

Just as current is induced from the primary into the secondary winding, it can also be induced into the iron core forming the transformer. This current causes a power loss in the core. This loss is about equal to the current in the core multiplied by the resistance of the core. It is referred to as an *eddy-current loss*. To minimize this power loss, the iron core is split into thin iron laminations or strips. Although all strips are placed side-by-side, they are insulated from each other by some sort of coating. This structure makes it more difficult for current to flow through the core than if the core were constructed from one solid piece of iron. Less current is therefore demanded from the primary and less power is wasted in the core.

One additional loss must be considered. The copper used for forming the coils of wire has some resistance. Power is wasted in these windings because of this *copper loss*.

After considering all these losses, you might think that transformers are very inefficient devices. But this is not the case. Power transformers used at power line frequencies, can have efficiencies of anywhere between 80% and 95%, so that only a small portion of the input power is actually lost. Efficiency of audio and high frequency transformers is usually much lower. Their efficiencies range between 50% and 85%.

Step Up and Step Down

Transformers come in many different shapes, forms, sizes, and circuits. When there are more turns forming the secondary winding than form the primary so that the voltage across the secondary is greater than the voltage across the primary, the component is referred to as a *step-up* transformer. Should the reverse be true and the voltage across the secondary winding is lower than the voltage across the primary, you have a *step-down* transformer. When voltages across both the secondary and primary windings are identical, it is usually referred to as an *isolating* transformer. This is because its function is usually to isolate the item being powered from being grounded by the supply furnishing the power.

Should there be more than one secondary winding on a transformer, it can perform both as a step-up and step-down device at the same time. A drawing of a transformer with three secondary windings is shown in Fig. 4-17. There is a 120-volt primary winding, a 170-volt secondary winding where the transformer performs its step-up functions, and two step-down windings to provide both 5 volts and 13 volts.

When using a transformer, be sure that the winding or windings being used, are capable of supplying the voltages and currents

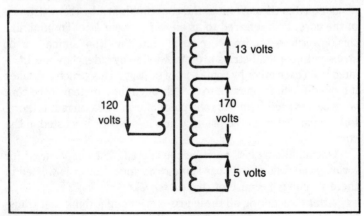

Fig. 4-17. Transformer with three secondary windings.

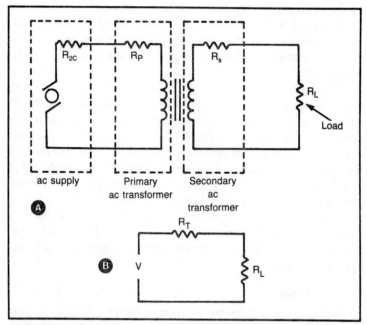

Fig. 4-18. A. Actual circuit. B. Equivalent circuit of arrangement in A. R_{ac} is resistance of power supply and is added to R_P, the resistance of the primary winding, before being reflected as R_T into the secondary. R_S is the resistance of the secondary winding, and R_L is the resistance of the load.

you need. If the load is such that you exceed the current limits of the winding as specified by the manufacturer of the transformer, it is liable to overheat and burn out. In this case, one or more of the windings may become charred or open.

Another factor to be aware of is that when voltage is fed to the primary of a transformer, the secondary winding appears as a voltage source to a load. In this case, the equivalent circuit of the transformer voltage source is the voltage across the secondary before any load is connected to it (open circuit voltage) in series with the resistance seen when looking into this winding. The voltage of the secondary winding to be used for this purpose is the voltage across that winding when the rated voltage is applied to the primary. The resistance of the equivalent circuit is approximately equal to the resistance, R_S, of the secondary winding added to the resistance seen in the secondary because of the resistance reflected from the primary. Thus, the actual resistance seen across the secondary, R_T, consists of these two factors. One is the resistance of the power supply plus the resistance of the primary winding, multiplied by the square of the turns ratio of the transformer. This is added to the second factor, the resistance of the secondary winding. All this is illustrated in Fig. 4-18A. A similar procedure was used to find the resistance seen in the secondary winding in the circuit in Fig. 4-16A.

Considering all this, the equivalent series circuit seen by a load across the secondary winding is the open-circuit voltage in series with a relatively complex resistance. An equivalent circuit of the actual situation is shown in Fig. 4-18B. If the voltage at the secondary terminals is to be in accordance with the voltage specified by the transformer manufacturer, the magnitude of R_L must be sufficiently large to minimize the voltage lost across R_T. Transformer specifications usually indicate the size this load must be in terms of the voltage that is available when a specific maximum current is demanded from the supply. The minimum load magnitude is determined by substituting numbers for the specified current and voltage into Equation 2-2.

Relative Polarity

Voltages at the terminals of the transformer windings vary in polarity with the voltage applied from the supply. During one-half of the ac cycle, the upper terminal of winding #1 of the transformer in Fig. 4-19 is positive with respect to the lower terminal. During the second half-cycle, the reverse is true. The upper terminal is now negative with respect to the lower terminal.

Because an ac voltage is induced into the secondary winding, the same situation exists there. Polarity across this winding alternates. Its positive and negative sequence is related to the sequence of polarities present across the primary winding. Dots are included at the terminals of some transformer schematics to indicate the relative polarities of the windings. The polarity of the terminals marked with dots is the same for all windings. In Fig. 4-19, during the time when the upper terminal of winding #1 is positive, the upper terminal of winding #2 is also positive. Another way of saying this is that voltages at the terminals with the dots are in phase with each other and out of phase with respect to the unmarked terminals of the various windings.

Now assume that the windings have a turns ratio of 3:1. If 120 volts is applied to the primary, 40 volts is across the secondary because the turns ratio is directly proportional to the voltage ratio. This is shown in Fig. 4-20A. Voltages at the upper terminals of both windings are in phase.

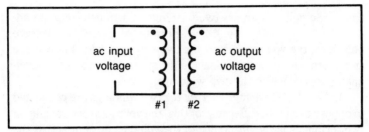

Fig. 4-19. Dots in schematic indicate relative polarities at terminals of windings.

In dc circuits, the voltages of two batteries add if they are connected in a series circuit with the positive terminal of one battery connected to the negative terminal of the second battery. The total voltage between the remaining two terminals is the sum of the individual voltages across each of the two batteries. The same is

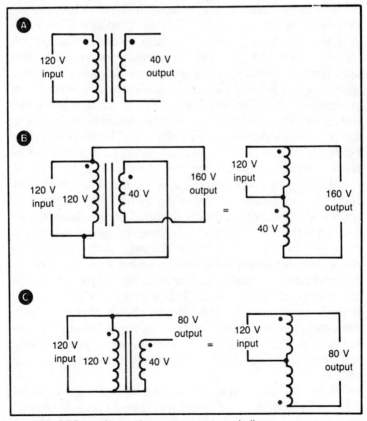

Fig. 4-20. Adding voltages that are across two windings.

true when voltages across two windings of a transformer are added. This is shown in Fig. 4-20B. Because of the relative voltage polarities, the voltages across the two windings add. The voltage at the remaining two terminals is the sum of the voltages developed across each of the two windings.

In Fig. 4-20C, voltages across the two windings are connected to oppose each other. Because of this, the voltage at the remaining two terminals is the difference of the voltages present across the windings.

The manner in which these voltages add or subtract can be illustrated by using phase diagrams. These are shown in Fig. 4-21. They consist of lines proportional in length to the magnitudes of voltages across the various windings. There is an arrow at the end of each line. The horizontal line with the arrow pointing to the right is the reference. This is shown in Fig. 4-21A. It represents the phase of the voltage across the input winding. It therefore forms an angle of $0°$ with the horizontal as if it were a radius in a circle. All windings that are in phase with the voltage across the primary winding are represented by horizontal lines with arrows pointing in this same direction to the right. If the windings are connected so that voltages add as in Fig. 4-20B, the lines with the arrows are drawn to indicate a sum. The final line is equal to the sum of the two shorter lines to indicate that the resulting voltage is larger than either of the two voltages producing this sum. This is shown in Fig. 4-21B, which illustrates the lines and arrows representing the individual voltages as well as the sum of these voltages.

Arrows are shown pointing in opposite directions if the phases of the two windings oppose each other by $180°$. The smaller voltage is subtracted from the larger voltage. The direction of the arrow on the line after this process has been completed determines the phase of the resulting voltage with respect to the input. Phases of the input and output voltages are identical if the magnitude of the difference (or output voltage) is to the right of the 0 point with the arrow pointing to the right. This is shown in Figs. 4-21C and D. In C, the two voltages with opposing phases, are shown. In D, one voltage has been subtracted from the other. Because the voltage being subtracted is less than the reference voltage, the phase of the final voltage is still $0°$ with respect to the reference or input voltage.

Should the line with the arrow pointing to the left be larger than the line with the arrow pointing to the right because the voltage across the secondary winding is greater than the voltage across

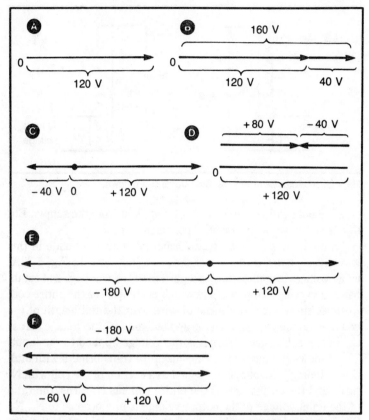

Fig. 4-21. Using vectors to add in-phase and out-of-phase voltages.

the primary, the resulting difference voltage would be indicated by a line with the arrow pointing to the left. The output or final sum would then be 180° out of phase with the input. This is shown in Figs. 4-21E and 4-21F. Here, there is 180 volts across the secondary of the transformer, rather than 40 volts. The final voltage across the two windings is 120 volts minus 180 volts for a difference of 60 volts. To indicate this difference, the arrow points to the left.

Autotransformers

The transformers discussed up till now have had two or more windings. But transformers can also be constructed from one tapped winding. Components with this structure are referred to as *autotransformers*. Their one big disadvantage is that individual wind-

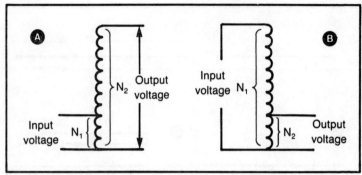

Fig. 4-22. Autotransformers. A. Step-up. B. Step-down.

ings are not available for isolating equipment from the supply. But this is not always an essential consideration.

Step-up or step-down transformers can also be made in this way. These two variations are shown in Fig. 4-22. In Fig. 4-22A, an ac voltage is applied across a portion of the winding, and establishes a varying magnetic field which encompasses the entire coil. Because there are more turns of wire over the full length of the coil than on the portion with the applied voltage, the voltage across the entire coil is greater than is the voltage applied to the input. This follows the turns ratio rules used for transformers with multiple windings. If voltage is applied to a section of the coil with N_1 turns and the output across the entire coil has N_2 turns, the ratio of the input voltage to the output voltage is proportional to N_1, N_2. Similarly the input to output current ratio is N_2/N_1, while the output to input resistance or impedance ratio is $(N_2/N_1)^2$.

The same logic and mathematical ratios apply to the step-down autotransformer in Fig. 4-22B. Now N_1 is greater than N_2 so that the output voltage is less than the input voltage.

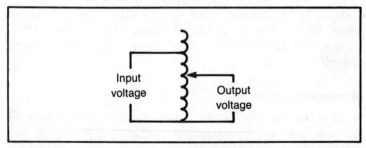

Fig. 4-23. Variable autotransformer or variac. Voltages can be stepped up or stepped down.

A *variac* is another type of autotransformer. Its output voltage can be varied from 0 volt to somewhat above the voltage applied to the input from the supply. A schematic of a variac is shown in Fig. 4-23. The arrow is a moving arm that slides over the coil. In different settings, it makes electrical contact with different turns on the winding. When it is set at the same point as is the tap to which the input voltage is applied, the output and input voltages are identical. When the moving arm or wiper is below the tap, the output voltage is less than the input. Should it be placed above the tap, the output voltage is greater than that at the input. The autotransformer combines the ideas illustrated in the two autotransformer arrangements shown in Fig. 4-22.

Chapter 5

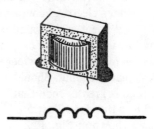

Passive Components
in AC Circuits

The resistances in a circuit affect the voltage and current in that circuit. This is true regardless of whether the applied power is ac or dc. But there are components that impede the flow of current only when ac voltage is applied. Two passive components of this type are capacitors and inductors.

INDUCTORS

Coils are turns of wire that resist the flow of ac current. This is because of the fluctuating magnetic field established when ac current flows in the turns of wire. Should only a dc voltage be placed across the inductor, the amount of dc current in the coil is determined by the resistance of the wire forming this component. The coil behaves like an ordinary resistor. But when ac current flows, there is also a reactance, X_L. *Reactance* is a term used to describe by how much a coil (or capacitor) opposes the flow of ac current. For an inductor, the reactance is determined by Equation 5-1.

$$X_L = \omega L \qquad \text{Equation 5-1}$$

Here, $\omega = 2\pi f$ or 6.28 times the frequency in Hz of the ac applied to the circuit, while L is the inductance of the coil.

If a coil and/or capacitor is in an ac circuit along with a resistor, the term *impedance*, rather than reactance or resistance, is applied. Impedance is a combination of a resistance and a reactance.

It indicates how the flow of ac current is impeded by the presence of these two types of components in one circuit.

The inductance shown in Equation 5-1 is a basic property of a coil. The unit of inductance is the *Henry*. A coil with a large number of turns has a large inductance and a big number of Henrys just as a large resistance would have a large number of ohms. An inductor with very few turns is given in terms of millihenrys (mH), or 1/1000 of a Henry, or in terms of microhenrys (μH), or 1/1,000,000 of a Henry.

Although the resistance of a component may be substituted directly into Equation 2-3 to determine the current in the device when a specific voltage is applied to it, this is not true of the inductance. Information about a coil's inductance, expressed in Henrys, must first be substituted into Equation 5-1 to determine the reactance of the coil. Once the reactance has been calculated, the current in the inductor can be determined using the ac equivalent of Ohm's law, where $e = iX_L$. Ac voltage, e, is the rms or root-mean-square voltage applied to the inductor, and i is the rms current flowing through the inductor when X_L is its reactance.

Equivalent Circuits

A pure inductor has only reactance to limit the flow of current. Real physical inductors also involve resistance. This is the actual resistance, R_S, of the wire forming the coil. An equivalent circuit of a real inductor is a resistor, R_S, in series with the inductor, L, as shown in Fig. 5-1. Here, R_S and L form what is known as the *series equivalent circuit* of an inductor. Its resistance is an important factor in determining the quality of Q or the coil.

When current flows through the inductor wired in a circuit, energy is applied to it by the power source. This energy causes a magnetic field to be formed around the coil. After the power

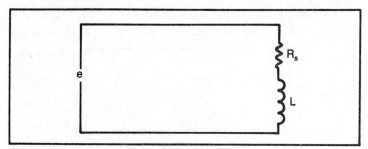

Fig. 5-1. Series equivalent circuit of an inductor.

source has been removed, all the energy should be returned from the coil to the circuit. None of the energy is dissipated by a perfect coil. But energy is dissipated by the resistance of the coil itself and the resistance of the circuit in which it is being used. Q relates the energy originally applied to the coil to the energy returned from the coil to the circuit. For the inductor itself, the Q or quality of the device when considering its series equivalent circuit is

$$Q = \frac{\omega L}{R_S} \qquad \text{Eq. 5-2}$$

Note that in this equation Q is inversely proportional to the magnitude of R_S. Consequently, if R_S is large, the Q of the coil is small.

A parallel resistor-inductor circuit can also be used to represent an inductor. This equivalent circuit is shown in Fig. 5-2. Here the resistance, R_P, is in parallel with the inductance. If this resistance is large, it has less effect on the overall circuit than it does if it is small. Consequently, it is desirable for R_P to be large—much larger than R_S in the series circuit. If the inductance in the two equivalent circuits is identical, the Q's of the two circuits are the same when R_P is equal to Q^2R_S. Thus, if R_S is 3 ohms in the series circuit and the Q of the circuit is 25, R_P must be equal to $(25^2)(3) = 1875$ ohms if the Q's of both circuits are to be identical. Actually, the Q of the parallel circuit is

$$Q = \frac{R_P}{\omega L} \qquad \text{Eq. 5-3}$$

It should be noted that the relationship between R_P and R_S given here is approximate. It is sufficiently accurate if Q is more than 3. If it is less than 3, then the relationship $R_P = (1 + Q^2)R_S$ should be used in its stead.

Through use of Equation 5-2, you can see that if Q = 25 in

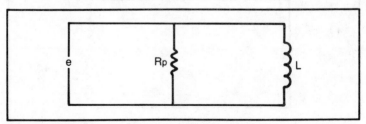

Fig. 5-2. Parallel equivalent circuit of an inductor.

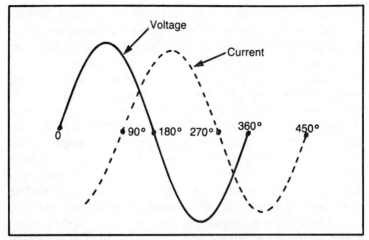

Fig. 5-3. Curves showing current in a coil lagging the applied voltage by 90°

the series equivalent circuit and R_S = 3 ohms, then ωL in this situation must be Q × R_S = 25 × 3 = 75 ohms. Substituting 1875 (calculated above) for R_P and 75 ohms for ωL into Equation 5-3, you will find that the Q of the parallel equivalent circuit is 1875/75 = 25. This is identical to the Q specified for the series equivalent circuit. It proves our original assumption that $R_P = Q^2 R_S$ when Q is greater than 3.

Relative Phases of Current and Voltage

When an ac voltage is applied across a resistor, cyclic current variations are *in phase* with the applied voltage. At the instant in the cycle when the voltage is at a peak, the current is also at a peak. This is shown in Fig. 4-6 where two in-phase sine waves are illustrated. In this case just disregard the sum.

The situation is different when an ac voltage is applied across an inductor. Voltage and current are not in phase. For a pure inductor, the current lags the voltage by 90° or $\pi/2$ radians. This is shown in Fig. 5-3. Here, the voltage cycle starts at 0° and goes through its complete sequence to 360°. The current cycle starts at 90° and is not complete until it reaches the 450° point.

Because current and voltage are out of phase, the real power dissipated (product of voltage and current) by an inductor is zero. This can be deduced from the curves in Fig. 5-3. At 0°, the instantaneous voltage is zero, while the instantaneous current is at a maximum. The product of the two, or power, is zero watts at this moment. The product is also zero at the 90° point where the voltage is at a peak and the current is at zero. The product of zero keeps repeating itself at 90° intervals for as long as the voltage and current cycles exist. In between these points, the negative product of current and voltage cancel the positive product to produce a zero average power over the entire cycle. This is obvious from the curves. Because the voltage is positive between 0° and 90° while the current is negative during this interval, the product of the positive and negative numbers, which equals power, is negative. The reverse situation exists between 180° and 270° where the voltage is negative and the current is positive. This product is also negative. Between 90° and 180°, both current and voltage are positive and between 270° and 360° both current and voltage are negative, so that both products are positive. Because the sum of the positive numbers indicating power in two quarters of the cycle is equal to the sum of the negative numbers in the two alternate quarters of the cycle, the sum of the positive and negative groups of numbers is zero. Thus, there is an average of zero watts of power over an entire cycle.

It is obvious from the equivalent circuits, that there is no pure inductor without an undesirable resistance. Because of the resistance in the overall circuit along with the inductance, the actual current does not lag the voltage by a full 90°. The angle of this lag is related to the relative magnitudes of the resistance and the reactance of the inductor, as well as to the frequency in question. This angle can be determined from a diagram showing the relative magnitudes of X_L and R as well as their relation to each other.

Phasor Diagrams

As in all circuits with components connected in series, the current flowing through both components in the series circuit in Fig. 5-1 is identical. The phase of the current in the inductor with respect to the supply voltage is identical to the phase of the current in the resistor with respect to the supply voltage. This is because the same current flows through both components in the circuit. Because current is identical throughout the entire circuit, this current is used as a reference in a diagram indicating how voltages are related to this current. As was discussed in Chapter 4, the reference is a horizontal line from a specific point. When the circle was discussed, this reference point was the center of the circle. Here too, it is considered as the "0" starting point. The line representing this reference current is shown in Fig. 5-4A. It is shown as a

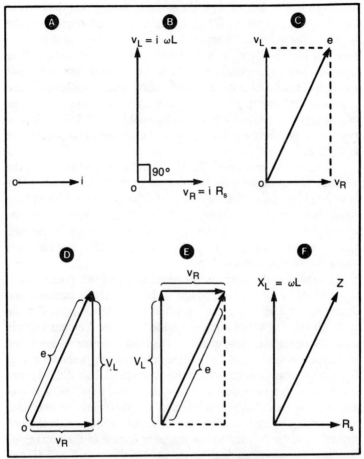

Fig. 5-4. Phasor drawings for series equivalent circuit in Fig. 5-1. A. Reference current. B. Voltages across R and L. Voltage across L leads the voltage across R by 90°. It also leads the current flowing in both component parts by the same 90°. C. Vector sum of v_R and v_L is equal to the supply voltage, e, D and E. Method of adding voltages. F. Adding the resistance, R, to the reactance, X_L, to get a total impedance, Z.

horizontal line with an arrow pointing to the right.

Voltage across the resistance section of the series equivalent circuit is in phase with the current, i, flowing through it. This voltage, v_R, is equal to iR_s. (Note that the voltage across the resistor is not in phase with the supply voltage.) Because i and v_R are in phase, the line representing the voltage developed across R is coincident in direction with the line representing the current.

Voltage across the inductance leads the circuit current by 90°

as shown in Fig. 5-3. (This is the same as saying that the current lags the voltage by 90°.) When the current in the inductor is at the 0° point, the voltage, v_L, across L is already 90° ahead of the current. In Fig. 5-4B, v_L is at the 90° point at this instant. The horizontal line representing the inductive current (and v_R), must rotate 90° in a counter-clockwise direction before reaching the line shown as representing the voltage, v_L. By that time, the voltage will have also rotated through another 90° to the 180° point, so that the voltage remains 90° ahead of the current in the inductance at all times.

Because the reactance, X_L, of the inductor section is ωL, the voltage, v_L across the inductive portion of the circuit is $i\omega L$. (This is derived by applying Equation 2-3 to the ac circuit.) The phase relationship of the voltage across the resistor, v_R, with the voltage across the inductor, v_L, is shown in Fig. 5-4B. Here it is assumed that v_L is twice the size of v_R. In ordinary circuits they can have any relative magnitudes.

The total voltage across the resistance and inductance, $v_R + v_L$, is equal to the supply voltage, e, because the inductance and resistance in a series circuit are wired across the supply. But the two individual voltages cannot be added directly. You must obtain a sum involving the phase angle, just as I did in the series-connected voltages shown in Fig. 4-7. This sum of v_R and v_L is shown in Fig. 5-4C. It is equal to e, the rms supply voltage. All we did here was complete a rectangle with v_R and v_L as two of its sides. The broken lines are equal in magnitude and are parallel to the two sides representing v_R and v_L. The sum or supply voltage e, is the line drawn from the 0 point to the opposite corner of the rectangle.

Another way of adding two voltages is shown in Fig. 5-4D and E. Here you add the quantities directly. In D, first draw the length representing v_R. Where v_R ends, v_L begins, so you should start the v_L line at the arrow on the line representing v_R. The total voltage, e, is the line from the beginning of v_R at the 0 point to the arrow of v_L. A similar procedure is shown in Fig. 5-4E where v_L is drawn first and then v_R starts at the arrow on the v_L line. In C, D, and E of the figure, the sums are all equal to voltage e.

Now note the diagram in Fig. 5-4F. It looks exactly like the one in Fig. 5-4C, but the lines have different labels. Instead of line v_R, it shows an R_S; instead of line v_L, it shows an X_L that is equal to the impedance of the inductance or equal to ωL; and instead of e there is a line labelled Z. Z is the reactance of the circuit and is equal to the vector sum of R_S and X_L.

The diagrams in F and C of Fig. 5-4 provide the same information. Because the current is identical in R_S and L, it is consequently identical to the current supplied by the voltage, e. The voltages across R_S and L, as well as the voltage, e, can be divided by this current without upsetting the proportional lengths or relative angles of lines representing these voltages. When the voltages shown in Fig. 5-4C are divided by the current, you are left with resistance, inductive reactance, and total circuit impedance. You know this from Ohm's law, Equation 2-2. Because of this, you are able to add the impedances in a series circuit in the same way you previously added the voltages.

Power Factor and Dissipation Factor

Let's pull the drawing out of Fig. 5-4F and redraw it as Fig. 5-5. But this time let us add information about the angles to the figure. Use R_S as the 0° reference. You do know that R_S and X_L are at 90° with respect to each other, but Z (as well as e in Fig. 5-4C) is at another angle, θ, with respect to the 0° reference. This angle, θ, is shown in the drawing. For convenience, I also labelled the remaining portion of the 90° right angle as $(90 - \theta)$ because it is 90° less the θ occupied by the angle between R_S and Z.

Now I can state several definitions used in the industry to specify inductors and more often to specify capacitors.

I have discussed the 90° phase difference between the current flowing in a pure inductor and the voltage across it, and indicated that the real power dissipated by an ideal inductor is zero. But you can still perform arithmetic and have a product of the voltage across the inductor with the current flowing through this component. It is not a power that is dissipated by the inductor so it is given the

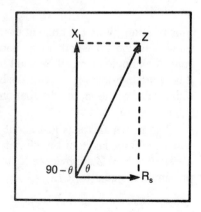

Fig. 5-5. Phasor diagram with information about leading and lagging angles.

name *reactive power*. It is not stated in units of watts, but in terms of *vars*.

The real power is dissipated by the resistance of the inductor or by any resistor that may be wired into the circuit in series with the inductor. Using Equations 3-4 and 3-5, this *active power* is the product of the current and v_R. This is identical to the product of the square of the current and R_S.

Using the real power as the $0°$ reference, the reactive power is at a $90°$ phase angle with respect to the real power. The power that the supply seems to provide is the vector sum of the active and reactive powers. This *apparent power* is the product of the current with e, or the product of the square of the current with Z.

Power factor is defined as the ratio of the actual power to the apparent power. Because the same current flows through R_S, X_L and Z, it is

$$\text{Power factor} = \frac{R_S}{Z} \cos \theta \qquad \textbf{Eq. 5-4}$$

The fact that it is related to the cosine of an angle can be seen from Equation 4-2.

Dissipation factor is the ratio of the amount of energy dissipated in the real inductor to the amount of energy stored in the inductor during one cycle. While the Q of an inductor shows how good it is, dissipation factor (and power factor) show how poor it is. Mathematically, dissipation factor is $1/Q$ or

$$\text{Dissipation factor} = \frac{R_S}{X_L} = \tan (90 - \theta) \qquad \textbf{Eq. 5-5}$$

This relationship to the tangent of the angle $90 - \theta$ can be derived using Equation 4-3. Another trigonometric relationship is the cotangent of an angle or *cot*. It is equal to $1/\tan$ of that angle so if tan A is an a/b in Equation 4-3, cot A is b/a. Then cot A = tan (90 – A). Using this definition, the dissipation factor can be stated as being equal to cot θ.

The power factor is just about identical to dissipation factor when the impedance of the circuit is very large compared to its resistance. Now Z approaches X_L in magnitude so that R_S/Z is approximately equal to R_S/X_L.

Calculating Impedance

Equation 4-4 relates the magnitudes of the sides of a right triangle to each other. In Fig. 5-5, R_S, ωL and Z form a right triangle. Using Equation 4-4 and Fig. 5-5, you can determine how the impedance of an inductor is related to its reactance and resistance. The formula is

$$Z^2 = R_S{}^2 + X_L{}^2 \text{ or } Z = \sqrt{R_S{}^2 + (\omega L)^2} \qquad \textbf{Eq. 5-6}$$

because $X_L = \omega L$. Of all the equations, this one and the one defining Q are the most frequently used when working with circuits that involve inductors.

Parallel Equivalent Circuit

In the analysis of the series equivalent circuit, the current was used as the reference or $0°$ angle because the same current flows through all components in the circuit. As for the $R_P = L$ parallel circuit in Fig. 5-2, the same rms voltage, e, is across the two components. You can logically conclude that the phasor diagram of a parallel circuit should use e as the reference. Current through all components can be determined from Ohm's law so that the current in the resistor is $i_R = e/R_p$, and current in the inductor is $i_L = e/\omega L$. But as always, the current in the inductor lags the current in the reactor by $90°$. All this is shown in Fig. 5-6A.

With the supply voltage as the reference, the current through

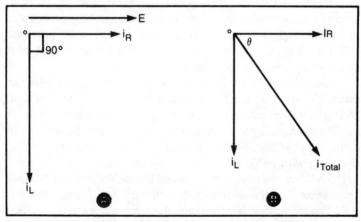

Fig. 5-6. Phasor diagram for parallel circuit. A. Current in inductor lags current in resistor by 90°. B. Vector addition of the current components.

the resistor is in phase with the voltage from the supply. Hence they are both drawn in the same direction starting at the zero point. Current in the inductor lags the reference voltage by 90°, so it is drawn at that angle with respect to the supply voltage. The line representing the current in the inductor must rotate 90° in a counter-clockwise direction before it can reach the e or 0° reference line. Hence as drawn, it shows i_L lagging the supply voltage by 90°.

The total current flowing through both components in the circuit, as well as the angle by which this total lags the supply voltage, is determined by adding currents in a manner similar to that previously used for adding the voltages of the series circuit. This is all shown in Fig. 5-6B. If the parallel equivalent circuit under consideration happens to be the parallel equivalent of the series circuit, angles, θ, are identical in both instances.

Identifying Inductors

Inductors come in many shapes and sizes. Some of the small inductors used in radio frequency circuits look just like low voltage resistors. The inductance of each component is determined by using the same color code, with minor exceptions, as was listed for resistors in Table 2-5. When inductance is less than 10 μH, the significance of each color band is as shown in Fig. 5-7A. Should the inductor be 10 μH or more, the significance of each band is as illustrated in Fig. 5-7B. It's easy to determine which sequence applies to a particular inductor. If there is a gold band located between two other bands, sequence A applies. If only the last band (marked "tolerance" in both drawings in the figure) is gold, or there is no gold band, or no fourth narrow band at all, then the sequence in Fig. 5-7B applies. In both cases, the inductance determined from the code is in microhenrys (μH).

When determining the inductance, disregard the wide silver band. Use the next band, the narrow band next to the wide one, to start the sequence for determining what the inductance is.

For example, assume you have a component with a narrow band sequence of yellow, gold, violet, gold. Using the drawings in Fig. 5-7 along with Table 2-5, you will find that this component is 4.7 μH, 5%. Here the gold band between the yellow and violet bands represents the decimal point between the 4 and 7, while the last gold band indicates that the component has a 5% tolerance, (See Fig. 5-7A). Should there be no last band so that the sequency of the narrow bands is yellow, gold, violet, the component is still 4.7

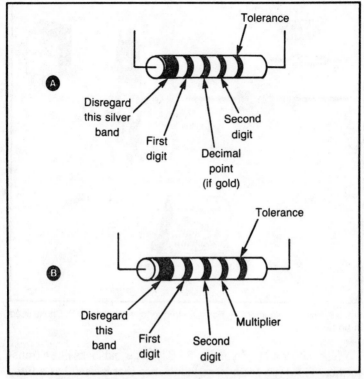

Fig. 5-7. Inductors that resemble 1/2-watt resistors. A. Component with less than 10 μH inductance. B. Component with 10 μH or higher inductance.

μH in inductance, but its tolerance is now 20%.

In another example, assume the color code sequence of the narrow bands is blue, gray, red, silver. Because there is no gold band between the first and third color, the drawing in Fig. 5-7B applies. The inductance of this component is 6800 μH and its tolerance is 10%.

Some typical physical forms of other inductors are shown in Fig. 5-8. Figure 5-8A shows samples of coils or chokes used primarily in radio-frequency circuits. One of these is shown with a screw protruding from its body. This screw is attached to some magnetic material. It rides up and down inside the coil form as the screw is turned. This maneuvering of the material is used as an adjustment to set the inductance of the coil to a desired value. Figure 5-8A shows an example of an inductor in a can. A screw adjustment may also be in this can to enable you to adjust the inductance to some required value.

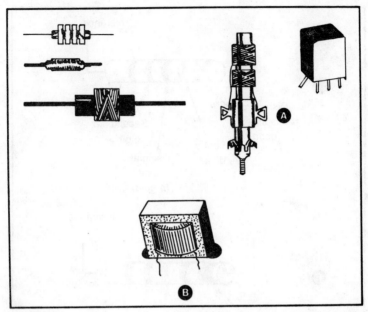

Fig. 5-8. Physical inductors. A. For radio-frequency applications. B. Choke used in power supplies.

The inductor shown in Fig. 5-8B is large and looks like a transformer. This type has a large inductance that is useful as a filter in a power-supply circuit. Although it is used in a supply designed to provide large quantities of power, it is seldom if ever found in the smaller power supplies found in radios, TV sets, and so on.

No color codes are used to indicate the inductance of any of the components shown in the drawing. The manufacturer usually supplies a separate piece of paper to indicate the inductance.

Examples

The reactance of an inductor can be determined using Equation 5-1. For example, assume you have a 1 mH (10^{-3}H) inductor. Its reactance at 100 Hz is $\omega L = 2\pi(100)(10^{-3}) = 0.628$ ohms. Should you want to use this inductor at 1 MHz, its reactance increases to $2\pi(10^6)(10^{-3}) = 6280$ ohms.

Another way of looking at this is by assuming you need an inductor to operate at 10 MHz, and it should have a reactance of 1000 ohms at that frequency. Using Equation 5-1, you find that you must get a component with an inductance of $L = X_L/\omega = 1000/6.28(10^6)$ $= 10^{-3}/6.28 = 0.16 \times 10^{-3}$ μH to satisfy your needs.

Should the individual inductors you have in your possession be the wrong values for use in the particular circuit you are working on, you can wire several inductors in series or in parallel to add to the inductance you need. The final inductance achieved using these methods is determined using procedures similar to those used when combining groups of resistors. The inductance of several inductors wired in series is the sum of the inductances of the individual components. If two inductors—one with an inductance of 10 mH and the second with an inductance of 20 mH—are wired in series, the inductance of the combination is 10 mH + 20 mH = 30 mH. Should these two inductors be wired in parallel, the resulting inductance of this circuit is (10 mH)(20 mH)/(10 mH + 20 mH) = 6.67 mH. Equations 3-2 and 3-3 are used here with the inductance being substituted for resistance.

Circuits in Figs. 5-1 and 5-2 were referred to as the equivalent circuits of actual real inductors. Methods used to analyze these circuits can be applied to circuits consisting of an inductor with a separate resistor. For example, assume you have an ideal component (0-ohm resistance) with a 10 mH inductance wired in series with a 10,000 ohm resistor. This combination is connected across a 120-volt source supplying 1 MHz. You want to determine the quantity of current flowing in this circuit, the voltage across each component, and the angle by which the current in these components lags the applied voltage. Initially, this circuit may seem as if it is only being used as an example of how the theory works. In actuality, this circuit is a very practical and commonly applied arrangement. Methods used to solve this problem can be applied to circuits used in practical applications.

At any rate, the solution is straightforward. The reactance of the 10 mH inductor at 1 MHz is $6.28(10^6)(10 \times 10^{-3})$ = 6.28 × 10^4 ohms = 62,800 ohms. Using Equation 5-6, the impedance, Z, of the overall circuit is $\sqrt{62,800^2 + 10,000^2}$ = 63,600 ohms. Applying Ohm's law, Equation 2-1, you will find that the total current in the circuit is 120V/63,600Ω = 1.89 × 10^{-3} A. Using Equation 2-3, the voltage across the resistor is (1.89 × 10^{-3} A)(10^4Ω) = 18.9 volts, and the voltage across the inductor is (1.89 × 10^{-3} A)(62.8 × 10^3Ω) = 118.69 volts. Note that if you add the voltages across L and R_S, the total voltage is 118.69 + 18.9 = 137.59 volts. This number is greater than the supply voltage. It seems impossible that the sum of the voltages across the two series-connected components in the circuit, should be greater than the voltage applied to the series circuit. In fact, it is impossible. The voltages cannot be added as shown, because the voltage across the inductor is out of phase with the voltage across the resistor. The phasor diagram illustrating this is in Fig. 5-9.

Using this drawing, you can determine by what angle, θ, v_R lags the supply voltage, e_{In}. Because $\tan \theta = v_L/v_R = 118.69/18.9 = 6.28$, $\theta = 81°$. Because the current and v_R are in the same direction or phase, the current lags the supply voltage, e, by this same 81° angle. Using Equation 4-4, you can see that with this phase lag the total voltage adds to the 120-volt supply or is very close to that value. The fact that the result of this computation may not be exactly 120 volts is due to some arithmetic tolerances in the calculations.

As a final example, assume that a 200,000 ohms (200×10^3 ohms) resistor is wired across (in parallel with) a perfect 15-mH inductor, and that they are both supplied 120 volts at 1 MHz. Now, you want to know the current in each component and the phase relationship of these currents to the supply voltage. The diagrams in Fig. 5-6 apply.

Using Equation 5-1, the reactance of the 15 mH inductor at 1 MHz is $6.28(10^6)(15 \times 10^{-3}) = 94.2 \times 10^3$ ohms. The current flowing through the inductor is $120/94.2 \times 10^3 = 1.27 \times 10^{-3}$ A. This current lags the supply voltage by 90°. The current in the resistor is $120/200 \times 10^3 = 0.6 \times 10^{-3}$ A. This current is in phase with the supply voltage. If the two currents were in phase, the total current required from the supply would be the direct sum of the currents in the two components. But they are out of phase. As

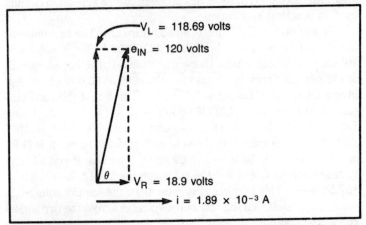

Fig. 5-9. Phasor diagram for problem involving series RL circuit. Supply voltage e = 120 V at 1 MHz, L = 10 mH, R = 10 kΩ.

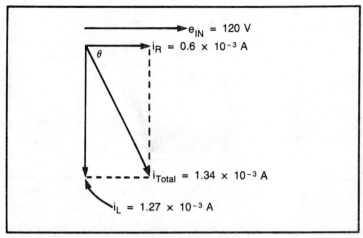

Fig. 5-10. Phasor diagram for problem involving parallel RL circuit. Supply voltage e = 120 V at 1 MHz, L = 15 mH, R = 200 kΩ.

a result, you must add the two currents vectorally as shown in Fig. 5-10. Using Equation 4-4, the total current is equal to

$$\sqrt{(1.2 \times 10^{-3})^2 + (0.6 \times 10^{-3})^2} - 1.34 \times 10^{-3} \text{ A}.$$

Because $\tan \theta = i_L/i_R = 1.27 \times 10^{-3}/0.6 \times 10^{-3} = 2.12$, $\theta = 64.7°$. Also, i_{Total} lags i_R by this angle. Since i_R is in phase with e_{In}, the total current supplied to the circuit lags the supply voltage by this same 64.7°. But it is important to remember that the current in the inductor itself lags the supply voltage by 90°, while the current in the resistor is in phase with the supply voltage and leads the current in the inductor by this same 90°.

CAPACITORS

The capacitor is a very frequently used circuit component. Next to the resistor, it is the most common electronic component. They are very common in ac circuits, especially in tuning and filtering applications. Capacitors are even more useful than inductors, and are simpler to understand.

Structure

A capacitor consists of two metal plates separated by some type of insulator. That insulator can be paper, mica, ceramic, and so on. It may even be air. This arrangement is shown in Fig. 5-11.

With no voltage applied to the two metal plates of a capacitor,

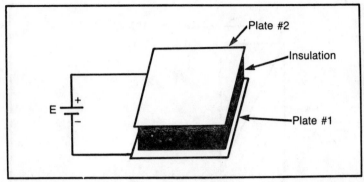

Fig. 5-11. Capacitor. All atoms in the insulator distorted when there is a charge on the plates.

the atoms in the insulating material retain their usual shapes. The electrons rotate in their usual orbits around the positive nucleus of the atom. When a dc voltage is applied to the plates, the orbits of the electrons in the atoms of the insulating material are warped. This is due to the field set up in this material by the presence of the positive battery voltage on one metal plate and the negative voltage on the second metal plate. The positive nuclei in the atoms are attracted to the negative metal plate, while the negative electrons in the orbits are attracted to the positive plate. This distorts the shape of the atom. This distortion remains even after the battery has been removed from the metal plates, because the individual charges on the plates have no where to go. The charge remains because the electrons cannot pass through the insulating material from one plate to the other. The capacitor is discharged only by connecting one metal plate to the other with a conductor (piece of wire) or by leakage through an imperfect insulator. After the capacitor has been discharged, the atoms in the insulator return to their normal shapes. But until this happens, the voltage across the plates remains the same as the voltage, e, originally supplied by the dc source.

Because of the insulator, capacitors do not conduct dc current. In fact, that is the reason the component retains the charge. However, the capacitor does conduct ac. The capacitor conducts current more readily at high frequencies than it does at low frequencies just the opposite of the inductor. The reactance of a capacitor, X_C, at any frequency, is

$$X_C = \frac{1}{\omega C}$$
Eq. 5-7

where C is the capacitance in farads.

The basic unit of capacitance is the farad, abbreviated F. The farad is a very big number. It is usually seen as $1/1,000,000 = 10^{-6}$ of a farad or in microfarads (μF). Even this is a large number, so many capacitors are made in $1/1,000,000$ of that size or in micro-microfarads ($\mu\mu$F) = 10^{-12} farads. But this is an old way of denoting 10^{-12} farads. It has been assigned the relatively new name, picofarads (pF). Therefore 1 pF = 1 $\mu\mu$F = 10^{-12} F. A more frequently used dimension between these two numbers is the nanofarad (nF), and is equal to 10^{-9} F.

Quality Factor

The insulators between plates of a capacitor are not perfect. There is always some minute amount of leakage. The equivalent circuit of a real capacitor with some leakage is a resistance in parallel with the capacitance. If this resistance, R_P, is present, as it normally is, the Q of the capacitor is

$$Q = \omega C R_P \qquad \text{Eq. 5-8}$$

This resistance is almost always very large so that the Q of a capacitor is very large. An equivalent circuit can also be formed where the resistance, R_S, is in series with the capacitance rather than in parallel with it. In this case, the resistance is extremely small. Here

$$Q = \frac{1}{\omega C R_S} \qquad \text{Eq. 5-9}$$

While the capacity, C, is the same in both equations and both equivalent circuits. R_P is approximately equal to $Q^2 R_S$.

Because Q is very large, it is seldom used to identify the quality of a capacitor. Instead, the dissipation factor shown as Equation 5-5 is used. Because it is equal to 1/Q, it is desirable that these numbers be as small as possible.

Dc Characteristics

Equations 5-7 through 5-9 define the ac characteristics and ac behavior of a capacitor. But when a dc voltage is applied to the component, it takes time for the capacitor to charge. Because there is no voltage across the capacitor at the instant the dc is applied, the capacitor behaves as if it had a resistance of zero ohms. Large quantities of current flow to charge the capacitor. As it continues charging, its apparent resistance increases while the current flow is reduced. Also the voltage across the component keeps increasing with time. After a period of time, usually several milliseconds, the capacitor is fully charged. Total applied voltage is now across the capacitor. If it is an ideal component, no current flows. It behaves as if it were presenting an infinite resistance to the dc supply.

The action of a capacitor in a dc circuit can be summarized by saying that at the outset, the voltage across the capacitor is zero and the maximum current the dc source can supply, is conducted to the component. After a short period of time has elapsed, the voltage across the capacitor is equal to that of the dc source and the current flow drops to zero.

Electrical Properties

Many different materials are used as insulators in capacitors. These materials are referred to as *dielectrics*. Each dielectric has a different characteristic. Capacitors with some dielectrics perform well at high frequencies, while others are good only at low frequencies. When a voltage greater than that specified by the manufacturer of a particular capacitor is applied to the metal plates, the dielectric may break down or short. When some types of dielectrics are used, the capacity of the component will change radically with temperature.

Electrolytic capacitors perform well up to about 1000 Hz but are usable in some applications up to 10 MHz. This type of capacitor uses aluminum oxide as the dielectric material. They are made with capacities up to 0.5 farads. The maximum voltage that may be applied to this type of capacitor, or its maximum rated voltage, is limited to 500. There are, of course, some electrolytic capacitors that cannot accept anything near this voltage. The voltage limit rating of a specific component is supplied by its manufacturer. Leakage of this type of component is high. Despite this, electrolytic capacitors are very useful as filters in power supplies and as coupling devices in transistor circuits.

Capacitors using tantalum as the dielectric are much smaller physically than the electrolytics. Their capacities are limited to about 1200 μF with breakdown peaks of 300 volts. They have the same high frequency limits as do the electrolytic devices.

Capacitors using paper or mylar as the dielectric material are quite useful up to 10 MHz. The maximum breakdown voltage is about 600 for both of these devices. They are made with capacities up to 10 μF, but capacitors with capacities above 0.5 μF are the exception rather than the rule. A variation on both types, metalized paper and metalized mylar, usually have a somewhat higher dissipation factor than do the ordinary paper and mylar capacitors, but are usually satisfactory. The later capacitors are frequently desirable because of their small size when compared to the size of regular paper and mylar components.

High frequency capacitors usually make use of ceramic, mica, or glass as the dielectric. They are usually limited to breakdown voltages below 1000 although there are certain ceramic capacitors that will take up to 40,000 volts. Although mica and glass capacitors are quite stable, the capacity of ceramic devices may vary considerably with temperature. Specifications are usually supplied to show the percentage of variation for the different capacitors.

There are many more different shapes and sizes of capacitors. Some of the more common ones are shown in Fig. 5-12.

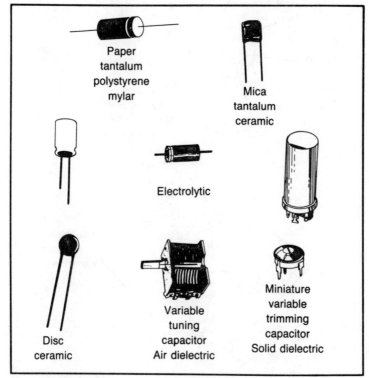

Fig. 5-12. Typical capacitors.

The size of the capacitor is usually stamped on the component. If an electrolytic is rated at 500 μF at 450 V, these numbers can usually be read directly from the inscription on the case of the component. In some instances, especially on disc ceramic and other small capacitors, its magnitude is noted as three digits. The first two digits are the significant figures, while the last one is a multiplier. So if 273 is imprinted, the capacity is 27,000 pF, which is equal to 0.027 μF.

When color dots are shown on the component to denote the capacity, the colors refer to the same numbers as for resistors. There is usually some arrow showing in which direction to read to colors. For example, a mica capacitor using six dots for the color code is shown in Fig. 5-13. So if the sequence of the top row of

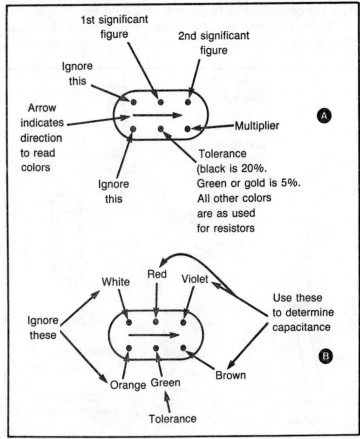

Fig. 5-13. A. Color code used for mica capacitors. B. Colors used in problem.

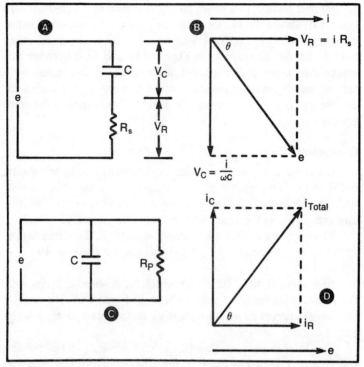

Fig. 5-14. Circuits and their phasor diagram. A. Series circuit. B. Phasor diagram for circuit in A. C. Parallel circuit. D. Phasor diagram for circuit in C.

colors, from the left to the right, is white, red, violet, and the colors in the bottom row are orange, green, brown, then the capacitance of the component is determined from the red, violet and brown colors, or 270 pF, and the tolerance is determined from the green, or 5%. The white and orange colors are ignored. This is shown in Fig. 5-13B.

Phasor Diagrams

A perfect capacitor has a characteristic that is just the opposite of a pure inductor. The primary characteristic that should be noted is that the current in a capacitor leads the applied voltage by 90°, while the current in an inductor lags the voltage by 90°. This is because ac current starts to flow through a capacitor at the instant that voltage is applied to its plates, but it takes time for the voltage to build up across the device.

A diagram of a voltage applied to a circuit consisting of a capacitor and resistor in a series circuit is shown in Fig. 5-14A. Figure 5-14B shows the phasor diagram. A similar representation of a voltage applied to a parallel resistor-capacitor combination, along with its phasor diagram, is in Figs. 5-14C and D. These are not unlike the diagrams in Figs. 5-4 and 5-6 for the inductors. Note that R_S and R_p may be resistors wired into the circuit with a capacitor, or they may represent the leakage resistance of the component.

Examples

Let us assume you have a mica capacitor such as the one shown in Fig. 5-13. The sequence of the top band of colors is white, brown, and red, while the sequence of its bottom band is orange, brown, and red. What is the reactance of this capacitor at 15 MHz?

The capacity is 1200 pF = 1200×10^{-12} F. The impedance, found by using Equation 5-7, is $1/6.28(15 \times 10^6)(1200 \times 10^{-12})$ = 0.113 ohms.

Now assume that a 0.2 ohm resistor, R_S is wired in series with the capacitor to form the circuit shown in Fig. 5-14A. If e = 0.02 volt, what current flows through the components and what is the phase angle?

The total impedance that this circuit presents to the voltage source is determined by using an Equation similar to Equation 5-6:

$$Z^2 = X_C^2 + R_S^2 \text{ or } Z = \sqrt{R^2 + \frac{1}{\omega C}^2} \qquad \textbf{Eq. 5-10}$$

Using this relationship, $Z = \sqrt{(0.2)^2 + (0.113)^2}$ = 0.23 ohms. If e is 0.02 volts, 0.02 volts/0.23 ohms = 0.087A = 87 mA must flow through the capacitor and resistor. The voltage across the resistor, using Equation 2-3, is 0.087×0.2 = 0.0174 volts. The voltage across the capacitor is 0.087×0.113 = 0.0098 volts. Because the current is not in phase with the applied voltage, the sum of the voltages across the capacitor and resistor, 0.0174 + 0.0098 = 0.0272, is greater than the applied 0.02 volt. This is identical to the case of the inductor and resistor in a series circuit where the total of the individual voltages is greater than the applied voltage. But this is just as impossible here as it was in the previous instance. The sum must be equal to the supply voltage. This sum is illogical because the voltages across the components are not added vectorally. When added in this fashion, using Equation 4-4 and shown in Fig. 5-15, e is equal to 0.02 volt. The current in the components

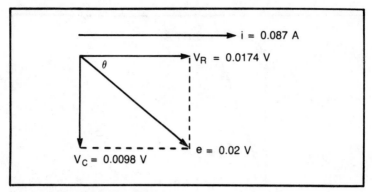

Fig. 5-15. Vector sum of voltages.

as well as v_R, lead the applied voltage by $\theta°$. Because tan θ = 0.0098/0.0174 = 0.5632, and θ = 29.4°.

The circuit has a Q because there is a resistor and a capacitor. Using Equation 5-9, Q is equal to $1/6.28(15 \times 10^6)(1200 \times 10^{-12})$ (0.2) = 0.023 so that the dissipation factor is 1/0.023 = 44.23. This would be very poor if there were only a capacitor in the circuit. But here you have a resistor wired in series with the capacitor so that 44.23 is the dissipation factor of the overall circuit and not solely the capacitor.

Instead of having a resistor connected in series with the capacitor, assume it is wired across the capacitor and assigned the symbol R_P. Assume R_P to equal R_S of the series circuit, or 0.2 ohm. If the same 0.02 volt, 15 MHz supply is used as in the previous example, the current through the resistor is $0.02V/0.2\Omega$ = 0.01A, and the current through the capacitor is $0.02V/0.113\Omega$ = 0.177A. The total current is determined from the vector sum of the individual currents as well as through use of Equation 4-4. This is shown in Fig. 5-16. It is $i_{Total} = \sqrt{0.01^2 + 0.177^2}$ = 0.1773 A. This is obviously less than the sum would have been had the phase angle not been taken into account. The phase angle is determined from the relationship tan θ = 0.177/0.01 = 17.7, so that θ = 86.77°.

Series and Parallel Combinations

At times it is convenient or even necessary to wire several capacitors in series or in parallel. If they are wired in parallel, the capacitance of the combination is equal to the sum of the capacitances of all capacitors in the circuit. Should three capacitors be wired in parallel, one with a capacitance of 5 μF, the second 8 μF and the

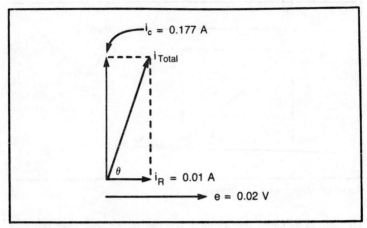

Fig. 5-16. Vector sum of currents.

third 10 μF, the capacitance of the parallel combination is 5 μF + 8 μF + 10 μF = 23 μF.

Should the capacitors be wired in a series circuit, an equation similar to 3-2 applies. Even though this equation was for a parallel resistor combination, it can be used for a series capacitor arrangement. Just substitute the C's or capacitances for the R's or resistances in the equation.

Let us say that the 5 μF, 8 μF, and 10 μF capacitors are in a series rather than a parallel circuit. The capacitance of only the 5 μF and 8 μF series combination is C_{s1} = (5 × 8)/(5 + 8) = 3.08 μF. This is the capacitance of only two of the three components connected in series, but a 10 μF capacitor is also in series with this combination. The capacitance of the entire combination is C_S = (3.08 × 10)/(3.08 + 10) = 2.36 μF. Note that this capacitance is less than that of any one of the capacitors in the series circuit and is certainly less than the capacitance of the parallel combination encompassing all these capacitors.

Power Dissipated

In all cases, the real power dissipated by an ideal capacitor or ideal inductor is zero. This is because of the 90° phase difference between the current through and the voltage across the component. Imaginary power is dissipated by the capacitor or inductor. It is the product of the current in the capacitor (or inductor) with the voltage across the component. The only real power dissipated is by the resistance in the circuit, which is equal to the product of the current flowing through R with the voltage that is across the resistor.

FILTERS

Up to now, it was always assumed that voltages applied to a circuit were either of the dc variety or at least a specific ac frequency. This is frequently the case. Power supplies usually provide voltages only at 60 Hz or else a pure dc similar to what is available from a battery.

But this is not the situation when audio signals or radio frequency (rf) signals are involved. Many different frequencies occur at the same time in any one group or *band*. The audio band consists of frequencies we can hear, covering the range from 20 Hz to 20,000 Hz. All these frequencies are available simultaneously for much of the time. In the radio broadcast band, almost all frequencies between 550 kHz and 1600 kHz are continuously present.

It is frequently desirable to reduce the voltage that is present at a specific frequency in a band or to reduce voltages that are present at frequencies above or below a particular frequency. This is most obvious on the audio band. In order to reduce voltages of the bass or low frequency portion of the audio band, a *bass* control on an amplifier is adjusted to accomplish this goal. In a similar manner, a *treble* control is used to reduce the ear-piercing high-frequency end of the audio band. In the rf band, it is always desirable to select the one frequency that is being broadcast by a specific radio station while eliminating frequencies broadcast by all other radio stations. Circuits that perform these functions are referred to as *filters*. Rf filters will be described in the discussion on *resonance*.

Signals in the band of frequencies are usually fed to amplifiers. These amplifiers produce gain and so amplify these input signals. After being amplified, a higher voltage or power at these signal frequencies is produced at the output of the amplifier than at the input. The amount by which the amplifier increases the input signals, is known as its "gain." Filters are designed to reduce the gain of a particular segment of these frequencies.

Audio filters are usually constructed using a series RC network. If you want to reduce the gain or relative signal levels of the low-frequency end of the band, you would use the circuit in Fig. 5-17A. Here capacitor C does not let the low frequencies pass as easily as the high frequencies. I noted this characteristic above when I discussed the fact that reactance is lower at high frequencies than at low frequencies. R and C form a voltage divider for the signals fed to the input of the network. Because X_C is greater at the low frequencies than it is at the high frequencies, less of the signal present at the input remains to be developed across R at these low frequencies after the signal has passed through C than at high frequencies. This can be deduced by using an equation similar to voltage divider Equation 3-1.

In Fig. 5-16B, the same voltage divider theory applies, only now, because the capacitor is across the output, the output is reduced as the frequency rises. Since X_C is reduced in magnitude with the increase in frequency, only a small portion of the input signal can be developed across the capacitor in the divider at the upper end of the band. Less signal is developed across C as the frequency rises.

The formula that is used to determine the reduction in the signal or gain at various frequencies in both circuits is

$$f_0 = \frac{1}{2\pi RC} \qquad \text{Eq. 5-11}$$

The output is reduced to 0.7 of its maximum magnitude or to 70% of maximum at frequency f_0. In Fig. 5-16B, the output drops to 0.45 of its maximum at $2f_0$ or at double f_0. It is reduced to one-quarter of its maximum at $4f_0$, one-eighth of its maximum at $8f_0$, and one-sixteenth of its maximum at $16f_0$, and so on. Thus, each time a specific frequency is doubled, the magnitude of the output signal is reduced by one-half.

Ideal audio amplifiers have identical gain at all frequencies between 20 Hz and 20 kHz. Assume you have an amplifier with a gain of 100. This indicates that when no filter is in the circuit, the input signal to the amplifier will be magnified 100 times by the circuit inside the amplifier before the signal reaches the output. Now the output signal is 100 times greater in magnitude than the input. A filter is used to reduce this gain at specific frequencies.

If you want to reduce the high frequency gain, you can add a network such as shown in Fig. 5-17B to the circuit. Let us design the network by making R equal to 10,000 ohms (or 10^4 ohms) and C equal to 0.0159 μF, which is identical to 1.59×10^{-8} F. Using Equation 5-11, you will find that f_0 is equal to $1/6.28(10^4)(1.59 \times 10^-) = 1000$ Hz. At this frequency, the gain has been reduced from 100 to $0.7 \times 100 = 70$. At double this frequency or 2000 Hz, the gain is now $0.45 \times 100 = 45$. At 4000 Hz, gain is $1/4 \times 100 = 25$; at 8000 Hz gain is $1/8 \times 100 = 12.5$; at 16,000 Hz gain is

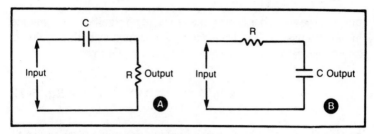

Fig. 5-17. Filters. A. High Pass. B. Low Pass.

$1/16 \times 100 = 6.25$. The next frequency that can be used to calculate the gain, 32,000 Hz, is outside the audio band and is consequently of no interest to us. The gain of frequencies between those indicated above can be estimated or interpolated from the known gains at the specific frequencies shown.

A similar situation exists when you want to reduce low-frequency gain using a filter circuit such as that shown in Fig. 5-17A. Here, gain is reduced to 0.707 of its maximum at f_0 Hz, to 0.45 of its maximum at $f_0/2$ Hz, to 0.25 of its maximum at $f_0/4$ Hz, to 0.125 of its maximum at $f_0/8$ Hz, and to 0.0625 of its maximum at $f_0/16$ Hz, and so on.

RESONANCE

Capacitors and inductors are usually combined in a circuit to do a specific job. They may be in a series circuit as shown in Fig. 5-18A or in a parallel circuit as in Fig. 5-18B. Both circuits have their individual characteristics. These characteristics vary with frequency. In both circuits R is the resistance of the coil. The resis-

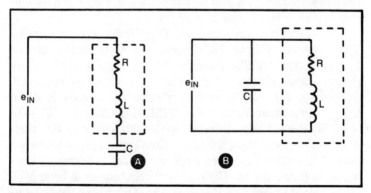

Fig. 5-18. Resonant circuits. A. Series resonant circuit. B. Parallel resonant circuit.

tance of the capacitor is presumed to be of no significant magnitude in our discussion. One frequency, f_O, common to both circuits, is known as the resonant frequency and is equal to

$$f_O = \frac{1}{2\pi \sqrt{LC}}$$ Eq. 5-12

Its significance will be described below. For now, let us return to our circuits. Both circuits can be analyzed using phasor diagrams.

Series Resonance

Three different frequency groups will be considered here. In one, the frequencies are so low that the reactance of the inductor is lower than the reactance of the capacitor. This is illustrated in Fig. 5-19A. Here, X_C is larger than X_L. They are drawn in such a manner as to show the proper phase relationships with respect to R. Because R is in phase with the current in the circuit, X_C and the voltage across the capacitor are shown as lagging this current by 90°, while X_L and the voltage across the inductor are shown as leading the current by 90°. These are the usual current-voltage phase relationships associated with inductors and capacitors.

In Fig. 5-19B, the magnitude of X_L is subtracted from X_C. This is quite simple because they are both on the same straight line but with arrows pointing in opposite directions. To determine the total impedance of the circuit, R must be added vectorally to this difference. The resulting line, $X_C - X_L$, added to R_S depicts the magnitude of the total impedance, Z, of the circuit at a relatively low frequency. It also depicts the angle, θ by which the current leads the input voltage, e_{In}.

A similar procedure can be employed if X_L is larger than X_C. This occurs at a relatively high frequency above resonant frequency f_O. The vector sum illustrates this situation in Fig. 5-19D. Here the impedance and supply voltage lead this current by an angle θ.

Finally, if the applied frequency is such as to make X_L equal to X_C, the situation shown in Fig. 5-19E exists. Because $X_C - X_L$ = 0, only R is left after one reactance is subtracted from the other. The voltage and current left in the circuit are in phase with each other. The total voltage is across R and is due to the current flowing through the circuit. The resonant frequency must be applied to the circuit to establish this condition. This frequency is determined using Equation 5-12. At this frequency, the impedance

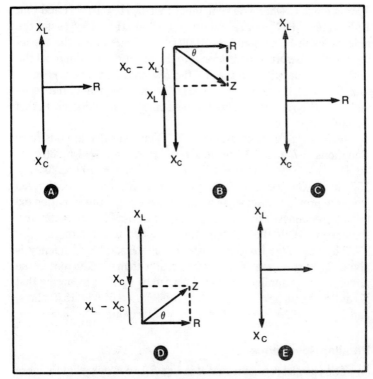

Fig. 5-19. Phasor diagram for series RLC circuit. A and B. Phasor diagrams when the frequency is less than f_O. C and D. Phasor diagram when the frequency is greater than f_O. E. Phasor diagram when the frequency is at resonance, f_O.

presented by this circuit is at a minimum and is equal to R. Because R is usually very small, the impedance presented to e_{In} by this circuit is very small at f_O. Note that f_O, the resonant frequency, is the dividing point between where the diagram in Fig. 5-19A applies and where the drawing in Fig. 5-19C applies. The impedance of the overall circuit is higher at the frequencies on either side of f_O than at f_O, as shown in Figs. 5-19B and D.

For example, assume you have a 1000 pF capacitor and a 500 μH inductor with a resistance of 1 ohm. At what frequency will the impedance of the circuit be 1 ohm? Will the current lead the voltage at 50 kHz or lag it? How about at 500 kHz?

First let's determine the resonant frequency. Substituting the data about the components into Equation 5-12,

$$f_O = 1/6.28(\sqrt{(500 \times 10^{-6})(1000 \times 10^{-12})}).$$

141

This is obvious when you remember that 1 μH = 10^{-6} H and that 1 pF = 10^{-12} F. You must multiply the 500 μH by 10^{-6} to get the inductance in Henrys and the 1000 pF by 10^{-12} to get the capacity in farads before substitution into the equation. By doing all the arithmetic, you will find that f_O, the resonant frequency, is equal to 225,193.24 Hz or about 225 kHz. At this frequency, the resistance of the circuit is equal to the resistance of the inductor. In this example, it is 1 ohm.

At 50 kHz, X_C is greater than X_L. You can calculate this from Equations 5-7 and 5-1, respectively. Here, X_C = 1/6.28(50 $\times$ 10^3)(10^{-9}) = 3185 ohms, and X_L = 6.38(50 $\times$ 10^3)(500 $\times$ 20^{-6}) = 157 ohms. The fact that X_C is greater than X_L can also be derived from the previous discussion which indicated that this situation exists at frequencies below resonance. Because X_C predominates, the current leads the voltage that is applied to the circuit.

The opposite situation exists at 500 kHz. This frequency is above f_O. In this case, X_L must be greater than X_C. Current in the circuit lags the applied voltage. This can be verified by noting that X_L at this frequency is 6.28(500 $\times$ 10^3)(500 $\times$ 10^{-6}) = 1570 ohms, and X_C = 1/6.28(500 $\times$ 10^3)(10^{-9}) = 319 ohms.

Parallel Resonance

The parallel resonant circuit in Fig. 5-18B is only slightly more difficult to analyze than is the series resonant circuit. The proper sequence for doing this is shown in Fig. 5-20.

Because this is a parallel circuit, e_{In} is used as the reference voltage. As I noted earlier, the current in the capacitor leads this voltage by 90°. This is shown in Fig. 5-20A.

Next consider the equivalent circuit of the inductor. It is a series circuit consisting of the inductance, L, and the resistance, R, of the component. As I noted earlier, the current, i_L, in a series circuit is used as the reference in the diagram. The voltage across the resistance, v_R, is in phase with the current. Voltage across the inductance, v_L, leads the current by 90°. The two voltages add vectorally. The sum is equal to the supply voltage, e_{In}. All this manipulation is shown in Fig. 5-20B.

But e_{In} in both figures is identical because the capacitor and the inductance/resistance combination are supplied power from the same source.

Now combine the sketch in B with that in A. Because e_{In} is the same in both diagrams, it is used as the reference. That means that the diagram in A remains as is, but in B it must be rotated θ degrees

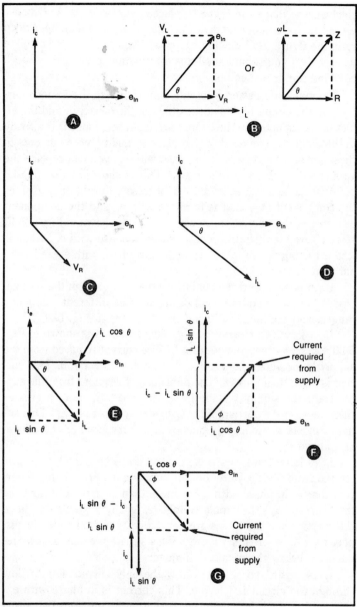

Fig. 5-20. Phasor diagrams for parallel RLC resonant circuit. **A.** Current in capacitor leads supply voltage by 90°. **B.** Vector sum of V_L and V_R and V_R is e_{In} because V_L leads i_L and V_R by 90°. **C.** Combination of diagrams in **A** and **B.** **D.** Replacing V_R with i_L. **E.** Horizontal and vertical component of i, are illustrated. **F.** The case where i_c is greater than $i_L \sin \theta$. **G.** The case where i_c $i_L \sin \theta$.

clockwise so that e_{In} in B can be placed over the e_{In} in A. Because v_L and v_R were added to give the total of e_{In}, you no longer need v_L in the figure. By knowing v_R and e_{In}, you can always calculate what v_L is if you should need this information at any time. So I'll combine the drawings but omit v_L.

To make life easier, you will want to utilize the various currents in the circuit to depict the effect of the capacitor and the effect of the inductor on the overall arrangement. Since i_L is already in the diagram, you need only replace v_R and e_{In} with currents in those phases. Because i_L is in phase with v_R, you can replace the v_R line with a line representing i_L. This is shown in Fig. 5-20D.

You can leave e_{In} as is, but it is necessary to split up i_L to show the component of i_L that is in phase with e_{In} and the component of i_L that is 180° out of phase with I_C. This is shown in Fig. 5-20E. Here $i_L \cos \theta$ is the portion of i_L that is in phase with e_{In}, while $i_L \sin\theta$ is the portion of i_L that is 90° out of phase with e_{In} and 180° out of phase with i_C.

I can now state three conditions that can exist in the parallel circuit. These conditions will occur in three different frequency ranges, and are analyzed using the diagrams in Fig. 5-20E.

i_C **is greater than** $i_L \sin \theta$. This situation is shown in Fig. 5-20F. $i_L \sin \theta$ is subtracted from I_C. The current required from the supply is the vector sum of $(i_C - i_L \sin \theta)$ with $i_L \cos \theta$. This current is more than $i_L \cos \theta$. The total current required from the supply leads the supply voltage by an angle $\phi°$. Since i_C is greater than the reactive portion of i_L, X_C is ideally less than X_L. This situation exists at the high-frequency end of the band or at frequencies above f_O.

i_C **is less than** $i_L \sin \theta$. As shown in Fig. 5-20G, i_C is subtracted from $i_L \sin \theta$. Here too, the supply must provide more current than is in phase with e_{In} or more than $i_L \cos \theta$. But now the total current lags the supply voltage by an angle labelled ϕ. Since the reactive portion of i_L is greater than i_C, you would ideally suspect that X_C is greater than X_L. This situation exists at low frequencies below the resonant frequency, f_O.

i_C **is equal to** $i_L \sin \theta$. Now the only current left to flow through the circuit is $i_L \cos \theta$. This current is in phase with e_{In}. Current $i_L \cos \theta$ is less than the current in the previous two situations. Because the same voltage is supplied to the circuit in all cases, the network must present a higher impedance in this situation than it did under conditions 1 and 2. This follows from Ohm's law which states that with a fixed voltage applied to a circuit, the impedance of that circuit is inversely proportional to the current flowing through it. With a fixed supply voltage, the impedance presented to the circuit must be higher with a lower current than with a higher current. This occurs at the resonant frequency.

Equation 5-12 is used to determine the resonant frequency of a parallel circuit, just as it was previously used for the series circuit. This is true with one exception. If R^2C/L is greater than one-tenth, then the following equation is a more accurate indication of the resonant frequency.

$$f_0 = \frac{1}{2\pi} \sqrt{\frac{1}{LC} - \frac{R}{L}^2} \qquad \text{Eq. 5-13}$$

A parallel resonant circuit presents its highest impedance to the supply at the resonant frequency. This impedance drops at frequencies above and below resonance. This is the opposite of the situation with the series circuit where the impedance was at a minimum at resonance and increased at frequencies on both sides of f_0.

An example will illustrate the methods used to determine the impedance of a parallel LC circuit such as that shown in Fig. 5-18B. Assume e_{In} is 100 volts. Using the components and data from the series resonant circuit example, $L = 500 \times 10^{-6}$ H, and $C = 1000 \times 10^{-12}$ F. But this time let $R = 100$ ohms. This is a pretty large number for R, but its use here is desirable for illustrating the calculation procedure. Using Equation 5-13,

$$f_0 = \frac{1}{6.28} \sqrt{\frac{1}{(500 \times 10^{-6})(1000 \times 10^{-12})} - \frac{100}{500 \times 10^{-6}}^2}$$

$$= \frac{1}{6.28} \sqrt{2 \times 10^{12} - (2 \times 10^5)^2}$$

$$= 222,817 \text{ Hz}$$

At 50 kHz, it was determined that $X_L = 157$ ohms and $X_C = 3185$ ohms, while at 500 kHz, $X_{L\$} = 1570$ ohms and $X_C = 319$ ohms. Now using this data, determine the current in the capacitor and its phase with respect to the voltage applied to the circuit. Also determine the current in the inductor and the phase of this current with respect to e_{In}, the magnitude of the current from the supply and its phase with respect to the applied voltage, the impedance of the circuit at the various frequencies, and the apparent, real, and reactive power in this circuit at the various frequencies. (Refer to Fig. 5-20 for the basic sequence used in calculating these factors.) Let's start working at 50,000 Hz. Information using data specified here is shown in Fig. 5-21.

At 50 kHz, $i_C = e_{In}/X_C = 100/3185 = 0.0314A = 31.4$ mA. It is at an angle of $90°$ with respect to e_{In}, as shown in Fig. 5-21A.

In the series inductor-resistor circuit, the relationship between the current in the LR combination and the voltages developed across the resistor and inductor are as shown in Fig. 5-21B. Again, e_{In} is the sum of these voltages, and i_L lags e_{In} by $\theta°$. The tangent of θ is equal to v_L/v_R. In Fig. 5-19B, the ratio of X_L to R was equal to the ratio of v_L to v_R. So you can rewrite this tangential relationship as $\tan \theta = X_L/X_R = 157/100 = 1.57$. Then $\theta = 57.5°$.

To determine i_L, you must first use Equation 5-6 to calculate the impedance of the RL circuit, which is $Z_L = \sqrt{R^2 + X_L^2} = \sqrt{100^2 + 157^2} = 186$ ohms. Then $i_L = e_{In}/Z_L = 100/186 = 538$ mA. A diagram showing the magnitude and phase relationships of currents in the inductor and capacitor, is shown in Fig. 5-21C.

Current i_L can be split into its horizontal and vertical components. This is shown in Fig. 5-21D along with I_C. The horizontal component of i_L is i_H. It is equal to $538 \cos 57.5° = 289$ mA. The vertical component of i_L, i_V, is $538 \sin 57.5° = 454$ mA. Subtracting I_C from i_V (454 mA $-$ 31.4 mA $=$ 422.6 mA) and taking i_H into account, the total current, i_{In}, provided by the supply is

$$\sqrt{289^2 + 422.6^2} = 512 \text{ mA}$$

This is shown in Fig. 5-21E. (It could just as readily have been summed up in the fashion shown in Fig. 5-21F.) Input current lags the input voltage by $\phi°$. This angle is determined from the equation $\tan \phi = 422.6/289$. Here, $\phi = 55.6°$.

Using Ohm's law, you will find that the impedance of the circuit is $e_{In}/i_{In} = 100$ V$/0.512$A $= 195$ ohms. The apparent power at this frequency is $e_{In}i_{In} = 100$ V $\times$ 0.512 A $= 51.2$ watts. But this is not the actual power demanded from the supply and dissipated by the circuit. This is because a portion of the apparent power is due to the product of a voltage and a current that are $90°$ out of phase with each other. Using the diagram in Fig. 5-21G, you can determine that the real power or power dissipated in the resistor is P $\times \cos \phi = 51.2 \cos 55.6° = 29$ watts, while the reactive or imaginary power across the circuit is $51.2 \sin 55.6° = 41.3$ watts.

146

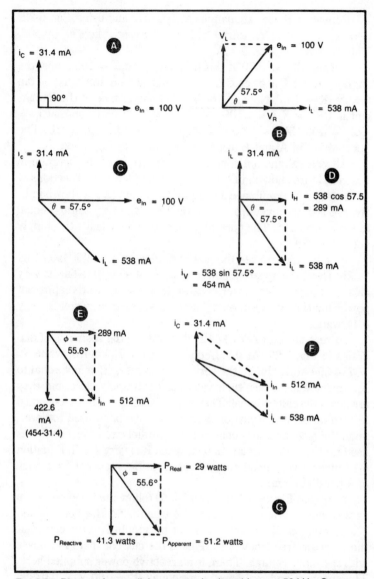

Fig. 5-21. Diagram for parallel-resonant-circuit problems at 50 kHz. Sequence shown in Fig. 5-20 is followed here with the addition of numbers.

Only the real power is actually being dissipated by the circuit because only here are the current and voltage in phase. Imaginary power is not drawn from the supply. The 51.2 watts is just an apparent power.

Repeating these calculations at 500 kHz and using Fig. 5-22, $i_C = e_{IN}/X_C = 100\ V/319\Omega = 313$ mA. Current i_C lags e_{In} by $90°$. This is shown in Fig. 5-22A.

Because $X_L = 1570$ at this frequency and $R = 100$ ohms, the angle by which i_L lags e_{In} can be determined from the equation tan $\theta = X_C/R = 15.7$. Now $\theta = 86.4°$. The impedance of the RL circuit is $\sqrt{R^2 + X_L^2} = \sqrt{100^2 + 1570^2} \approx 1573$ ohms so that $i_L = V/Z_L = 100/1573 = 63.6$ mA. This is shown in Fig. 5-22B. The information in A and B are combined to form Fig. 5-22C.

All vertical and horizontal current portions of the diagram in Fig. 5-22C are shown in D. The vertical and horizontal sections of the currents are combined and summed vectorally, as shown in E. Here, ϕ is determined from the ratio 249.5/4, for this ratio is equal to tan ϕ. ϕ is $89°$. The current at this frequency leads the supply voltage by $89°$.

Finally, you can see the power dissipated at 500 kHz in Fig. 5-22F. Here, the apparent power, $e_{In}i_{In} = 24.95$ watts. This is very close to the reactive power because this is equal to the apparent power multiplied by sin $89°$. The real power is very small—only 0.44 watts.

At resonance of 222,817 Hz, you will use the sequence of diagrams in Fig. 5-23. At this frequency, $X_L = 700$ ohms, while $X_C = 714.29$ ohms. Unlike the series circuit case, X_L is not equal to X_C. In some instances, their values are extremely close, but they are never absolutely equal. They would be identical if R were equal to zero ohms, and Equation 5-12 could have been used to determine the resonance frequency of the parallel circuit just as it was used for the series circuit. In fact, when R is very small, Equation 5-12 gives a very good approximation to the resonant frequency of a parallel circuit.

Note that Fig. 5-22A through 5-22D follow the same sequence as was shown originally in Figs. 5-21 and 5-22. But here, i_V and i_C are equal so all current that must be supplied to the circuit at the resonant frequency is i_H. This is the current that is in phase with the input voltage. There is no reactive power supplied here. The apparent and real powers are equal—$P = 100\ V \times 0.01995\ A = 1.995$ watts.

The impedance provided by this circuit at resonance can be determined by applying Ohm's law. It is the input voltage divided by the total current, i_{In}. At resonance, it is 100 V/19.95 mA = 5012.5 ohms. At 50 kHz is 100 V/512 mA = 198.3 ohms, and at 500 kHz it is 100 V/249.5 mA = 400.8 ohms. Note how much higher

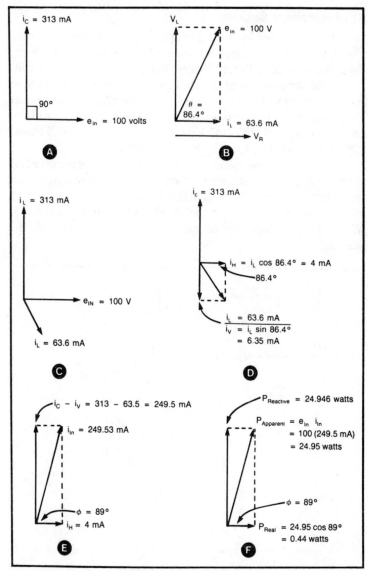

Fig. 5-22. Diagrams for parallel-resonant-circuit problem at 500 kHz.

the impedance is at resonance than at the other frequencies. The important thing to remember here is that for parallel resonant circuits, the impedance presented by the circuit is at a maximum at the resonant frequency while for series resonant circuits the impedance is at a minimum at that frequency.

Q of Resonant Circuits

Q is determined by the losses in a circuit. In resonant circuits, the only loss of significance is the resistance, R, of the coil. Therefore, the Q of the coil is effectively the Q of the resonant circuit.

In the series resonant circuit, R was specified in the example as being equal to 1 ohm. Here, as in the parallel circuit, the coil could have had any resistance. At resonance, the reactance of the inductor is $X_L = 6.28(225,193 \text{ Hz})(500 \times 10^{-6} \text{ H}) = 707.5$ ohms. Using Equation 5-2, $Q = X_L/R = 707.5/1 = 707.5$.

At resonance, the voltages across both the series-connected inductor and capacitor are higher than the voltage provided by the supply. In both cases, the magnitude of the voltage across each of the components, is equal to the supply voltage multiplied by Q. When Q is very high, the voltage across each component is also very high.

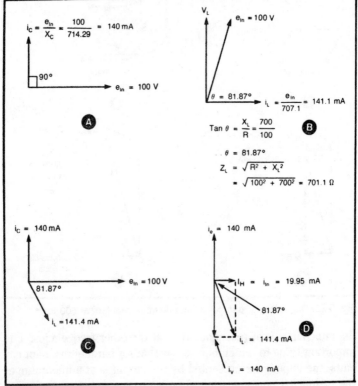

Fig. 5-23. Diagrams for parallel-resonant-circuit problems at the resonant frequency, f_o.

150

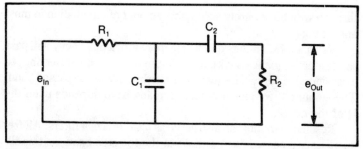

Fig. 5-24. Bandpass RC filter.

In the parallel resonant circuit, Q is determined in the same manner as in the series circuit. In our example, R was chosen to equal 100 ohms and X_L at the resonant frequency of 222,817 Hz was 700 ohms. Q in this case is 700/100 = 7. The resistance presented to the supply by this circuit at the resonant frequency is approximately equal to Q^2R. In this example, it would be $7^2 \times 100 = 4900$ ohms. This is quite close to the resistance of 5012.5 calculated above. This difference is small enough so as not to be of any practical significance.

At resonance, the currents in both the inductor and capacitor in the parallel arrangement are close to being equal to Q times the input current, i_{In}, demanded by the entire circuit. In our example, the input current is 19.95 mA. Multiply this by 7, the Q of the circuit. The product, 140 mA, is the current flowing through each component. It is a close approximation to data found in our previous calculations.

Filters

RC filters are used to attenuate either the upper end or the lower end of a band of frequencies. Two filters can be combined into one circuit to let only one group of frequencies in the band pass. A circuit of this type is shown in Fig. 5-24. Here the R_1-C_1 network rejects the upper of high frequency end of the band, while the R_2-C_2 network rejects the lower frequency end of the band. Only frequencies between the two rejected groups can pass with a minimum of attenuation.

RC networks are often used in low-frequency applications to perform this function and are known as *bandpass filters*. The actual passband is usually quite wide when these circuits are used and the *rolloff*, or the amount of frequency attenuation on either side of these frequencies, is low. Even the rejected frequencies pass through with but a barely acceptable amount of attenuation in most applications.

Resonant LC circuits do this job more efficiently. They will pass the center frequencies with a minimum of loss while rejecting, to a considerable degree, frequencies on either side of the passed band. The amount of frequencies passed in any band depends upon the Q of the circuit.

Resonant circuits can also serve as *band-rejection* filters. All frequencies are passed except those in the rejected band, so this filter behaves just the opposite of a *bandpass* filter. Basic diagrams of circuits for both types of filters using series and parallel resonant circuits, are shown in Fig. 5-25.

The resonant circuit in Fig. 5-25A has a low impedance at or near the resonant frequency and has a high impedance at all other frequencies. Because of its low impedance, the circuit lets voltages at frequencies in the vicinity of F_O pass rather freely from the input to the output. At the other frequencies, the high impedance of the circuit impedes the flow of signal. If the Q of the circuit is high, this circuit behaves as an ideal bandpass filter. It will perform its function only if a resistor or load is connected across the e_{Out} terminals.

The LC circuit in Fig. 5-25C shunts frequencies at and around f_O to ground. This is because of the circuit's low impedance when presented with frequencies in this group. Because the impedance of the circuit is high at frequencies outside of this narrow band, it has no or little effect on signals at the remaining frequencies. This arrangement consequently behaves as an ideal band-rejection filter. Resistor R is in the circuit to isolate e_{Out} from e_{In}.

The parallel-resonant circuits in Figs. 5-25B and D are also ideal as filters. This is due to the high impedance presented by the LC circuit at and around the resonant frequency and the low impedance presented by the circuit at frequencies above and below f_O. In Fig. 5-25B, the resonant circuit shunts signals at all frequencies except those near and at f_O while presenting a high impedance to frequencies at and near f_O. Because it lets only a narrow band of frequencies pass from e_{In} to e_{Out}, this circuit is an excellent bandpass filter. Resistor R serves the same function here as it did in the circuit in C of the figure.

The parallel-resonant circuit shown in Fig. 5-25D, does not let signals pass from e_{In} to e_{Out} at and around f_O because it presents a high impedance to signals in this group of frequencies. Signals at all other frequencies pass rather freely. Consequently, the reso-

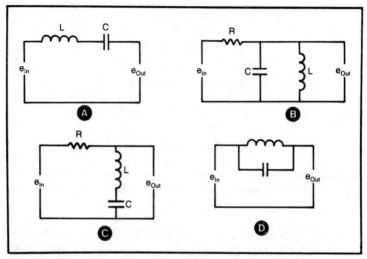

Fig. 5-25. Circuits in A and B are bandpass filters, while circuits in C and D are band-elimination or band-rejection filters.

nant network in this circuit arrangement behaves as a band-rejection filter. As in Fig. 5-25A, a load must be wired across the e_{Out} terminals if the circuit is to function properly.

A frequency-response curve due to the action of a bandpass filter in a circuit is shown in Fig. 5-26. The center frequency, f_O, is passed most readily by the resonant circuit. Output at frequency f_L below f_O and at f_H above f_O is 0.7 of the maximum output possible from the circuit, or is 0.7 e_{Max}. When the level of the signal has dropped to 0.7 e_{Max}, the gain of the circuit is considered as not having dropped to a noticeable degree. Therefore, in Fig. 5-26, the portion of the band between f_L and f_H is considered as the portion of the band passed by the components involving the resonant circuit. The gain at frequencies below f_L and above f_H is relatively low because they are outside of this passband.

The size of this passband is related to the Q of the circuit. It can be determined from the equation

$$f_H - f_L = \frac{f_O}{Q} \qquad \textbf{Eq. 5-14}$$

where all quantities are identical to those indicated in Fig. 5-26. The Q of the resonant circuit is substituted for the Q in the equation.

For example, consider the series-resonant circuit discussed earlier. Because f_O = 225,193 Hz and Q = 707.5, f_H-f_L =

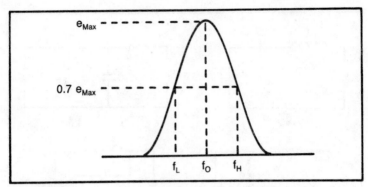

Fig. 5-26. Curve that can be used to determine the bandwidth and Q of a resonant circuit.

225,193/707.5 = 318 Hz. If f_L is as much below f_O as f_H is above it, f_H is 225,193 + 318/2 = 225, 352 Hz, and f_L is 225,193 − 318/2 = 225,034 Hz. Therefore, the bandwidth of this circuit is from 225, 034 to 225, 352 Hz. This is an extremely narrow band around a 225, 193 center frequency. If you want to widen it, you would have to add a resistor in series with L to reduce Q. If you should want to make the passband narrower for some strange reason, you would then need a different coil with a higher Q than 707.5. This coil must then have less than a one-ohm resistance.

Next consider the parallel-resonant circuit where f_O is equal to 222,817 Hz and Q = 7. Now $f_H - f_L$ = 222, 817/7 = 31,831 Hz so that f_H = 222,817 + 31, 831/2 = 238, 733 Hz and f_L = 222,817 − 31, 831/2 = 206,902 Hz. The passed band from 206, 902 Hz to 238,733 Hz is extremely wide when it is considered around a center frequency of 222,817. It can be reduced only by replacing the coil with one exhibiting a lower resistance and a higher Q.

A Q of about 100 is quite common for many of the coils. If you are working at frequencies around 200,000 Hz, the bandwidth or $f_H - f_L$ would then be 200,000/100 = 2,000 Hz.

Let us finally determine what coil to specify if the bandwidth should be 10,000 Hz with 455,000 Hz as the center frequency. Substituting the numbers into Equation 5-14, you will find that 10,000 = 455,000/Q. Solving for Q, it must be equal to 45.5. This is a requirement for a coil that may be used in one section of an ordinary radio.

Chapter 6

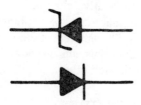

Diodes

Passive components can be defined as devices that do not amplify signals or do not electronically switch voltages or currents on and off. Resistors, capacitors, and industors fall into this category. Up to this point, I have discussed only components that satisfy the requirements of this classification.

Active components have other capabilities. Some of them may be used as amplifiers of voltage or current. Others can perform as electronic switches. One such device, the diode, will be discussed here. Devices in this group are composed of semiconductor slabs.

SEMICONDUCTOR MATERIALS

In previous chapters, I noted that there are two basic types of materials—conductors and insulators. Copper that is conductors and insulators. Copper that is used to form wire is a good conductor. The material around the copper wire, usually cotton or plastic, is an insulator. Should two wires be touching, this plastic insulator keeps one conductor isolated from the other.

But there are also a materials referred to as *semiconductors*. Materials of this type, such as silicon or germanium, are frequently used to produce electronic components. Pure germanium and silicon are both fair insulators. They are mixed with other materials or impurities to create truly useful semiconductors.

When the basic element, germanium or silicon, is combined with an impurity such as arsenic or phosphorous, an *n-type* semiconductor is formed by the combination. This new material has some mobile electrons, so it exhibits a negative charge.

A *p-type* material can be formed if an impurity such as aluminum or indium is added to the basic element. Now there is a shortage of electrons or an excess of mobile *holes* in the combination. This combination has a positive charge.

These two types of materials are combined to form different semiconductor devices. The simplest of these is the *junction diode*. It is composed of two materials, one n-type and one p-type, placed in intimate contact with each other.

JUNCTION DIODE STRUCTURE

Two semiconductor slabs that have been arranged to form a diode, are shown in Fig. 6-1. It's obvious that when you have a highly negative material adjacent to a highly positive material the positive material will attract the negative electrons and the negative material will attract the positive holes. By the time this interchange has been completed, an area is formed near the junction of the materials. The portion of the area located at the edge of the n-type material becomes positively charged. Negative charges migrate to the section of the area near the edge of the p-type material. The areas containing these two groups of charges, are known as *depletion regions*. This is shown in Fig. 6-2. The depth that the charges penetrate into the adjacent regions depends upon the magnitude of the charges in the region as well as on the number of charges required before the build-up is sufficient to prevent additional charges from crossing the junction. These additional charges are repelled by charges already in the regions.

If a battery is connected across the diode, the effect on the circuit depends upon the orientation of the battery with respect to the

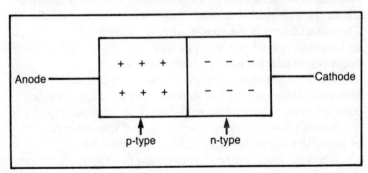

Fig. 6-1. Basic structure of the junction diode.

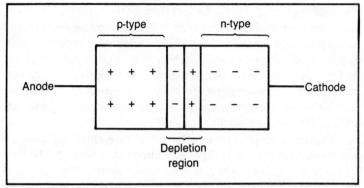

Fig. 6-2. Depletion region is formed at the junction of the n-type and p-type semiconductors.

two diode terminals as well as on the magnitude of the battery voltage.

When the positive terminals of the battery are connected to the n-type material and the negative to the p-type material, the effective width of the depletion region is increased. No current can flow from one terminal of the battery to the other.

By reversing the connections to the battery, the charge in the depletion region is neutralized. There is no longer a barrier for the current from the battery to flow through the diode.

The schematic symbol for a diode is shown in Fig. 6-3A. The p-type material is labelled the *anode,* while the n-type material is the *cathode.* In Fig. 6-3B, the positive terminal of the battery is connected to the anode, and the negative terminal is connected to the cathode. Because the anode is positive with respect to the cathode, the diode conducts current. It conducts electron current from

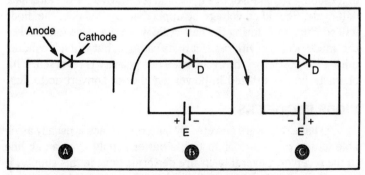

Fig. 6-3. A. Schematic symbol of diode. B. Forward-biased diode. C. Reverse-biased diode.

157

the negative terminal of the battery through the cathode, the anode, and then back to the positive terminal. But note that one angle of the anode points to the cathode. This angle is like an arrow pointing in the direction opposite to the width electron current flows. In fact, the arrow on the line to indicate the direction in which current is flowing also points in the same direction as does the arrow in the symbol. This arrow also points in a direction opposite to that of the electron current flow.

Explanations of semiconductor circuits are simplified if one assumes that current really flows in a direction opposite to the flow of electrons. This is known as *conventional current.* This theoretical current flows from the positive terminal of the battery through the components in the circuit to the negative battery terminal. If a diode is connected to the battery as it is in Fig. 6-3B, conventional current flows as shown by the arrow from the positive battery terminal through the anode to the cathode and then back to the negative battery terminal. Current will flow only if the battery is connected in such a fashion that the *arrow* in the diode symbol is positive with respect to the *straight line* cathode of that symbol. The arrow of the symbol actually indicates the direction in which conventional current can flow through the diode. Thus, the arrow or anode must be positive with respect to the cathode if conventional current can flow through the diode. Thus, the arrow or anode must be positive with respect to the cathode if conventional current is to flow. It will flow in the direction pointed to by the arrow, from the anode to the cathode.

Current will not flow when the anode is negative with respect to the cathode. This is shown in Fig. 6-3C.

One ac voltage is positive with respect to the second terminal of its supply for one-half of a cycle and is negative for the alternate half-cycle. Should ac voltage be applied across a diode, the diode will conduct only during the half-cycle when the ac voltage makes the anode positive with respect to its cathode, but will not conduct during the portion of the cycle when this polarity is reversed. This characteristic is utilized in power supplies to convert ac to dc.

DIODE RECTIFIERS

Portable radios are powered by batteries. A jack is usually available on the radio for connecting a power supply from an ac line to the receiver. Several tasks are performed inside this supply. It reduces the 120 volts ac available from the power line to a lower ac voltage suitable for use as a power source for the radio. It then converts the ac to pulses of dc. Finally, a filter is added to the circuit to smooth the ripple due to the voltage fluctuations in each ac cycle to shape it into a smooth dc voltage. By plugging the output connector on the supply into the power jack on the radio, the dc from the supply is fed to the radio to power it. At the same time, the battery inside the radio is disconnected from the circuit. When the battery is disconnected, the ac to dc converter in the supply takes over.

A circuit involving a power transformer and diode is shown in Fig. 6-4. Load resistor R_L represents the load a radio would place on the supply. If the electric company supplies 120 volts ac and you only need 9 volts ac, then a transformer can be used to accomplish this voltage reduction. Using Equation 4-8, you will find that the transformer must have a turns ratio that is equal to the required voltage ratio, or 120:9 = 13.33:1. This is because the turns ratio is proportional to the voltage ratio. If you want to buy the transformer, you would simply indicate to the salesman that you need a component that will accept 120 volts ac on its primary winding and give you an output of 9 volts ac on its secondary winding. You should also tell him just about what current the power-supply circuit must provide so that he will sell you the smallest and least expensive transformer that is capable of doing the job without overheating.

The ac voltage that you hear about most of the time is the average or root-mean-square value. The electric company supplies us with 120 volts rms. The transformer has 9 volts rms across its secondary winding. Using Equation 4-6, you can find that the peak voltage across the secondary winding is 9 volts times $\sqrt{2}$ or 12.7 volts. Since each cycle of the sine wave looks like that shown in Fig. 4-12, the positive peak voltage across the secondary winding,

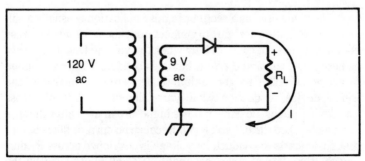

Fig. 6-4. Circuit used to reduce the ac supply voltage to useful levels and convert the ac to a pulsating-dc voltage.

$+E_{Peak}$, is $+12.7$ volts and the negative peak, $-E_{Peak}$, is -12.7 volts.

Assume that the lower terminal of the 9-volt winding of the transformer is used as the reference point. During the portion of the cycle when the upper terminal is positive with respect to the lower terminal, the anode of the diode is biased positive with respect to the cathode and the diode conducts. It conducts conventional current from the positive terminal of the 9-volt winding through the diode to the load resistor, R_L, and to the negative terminal of the secondary winding. Because of the direction in which the conventional current is flowing, the upper terminal of R_L is positive with respect to its lower terminal. The 12.7-volt positive half-cycle sinusoidal pulse is developed across R_L.

On the alternate half-cycle, the upper terminal of the 9-volt transformer winding varies down to $-E_{Peak}$ or is negative 12.7 volts with respect to its lower terminal. Because of this polarity, the anode of the diode is negative with respect to the cathode so the diode does not conduct. The voltage across R_L is zero during this half-cycle because no current flows through it.

These events keep repeating for each subsequent cycle. If the supply has a frequency of 60 Hz, then these events repeat sixty times each second. The end result of several of these cycles, is shown in Fig. 6-5. One-half of a sine wave is developed across R_L during one half of each cycle and a zero volt level is available during the alternate half of each cycle.

Because the diode is a good conductor when the anode is positive with respect to the cathode, it might occur to you that the diode has zero resistance. As a result of this, none of the dc voltage is developed across the diode and all of it is instead developed across R_L. Thus, E_{Peak} across R_L should be 12.7 volts. But this is not absolutely true.

The diode used as a rectifier in this application is usually a silicon device. It does not start to conduct until about 0.7 volt is across it. Thus it does not conduct fully when the voltage from the supply is between 0 volt and 0.7 volt, but does conduct when the voltage is above 0.7 volt. (If a germanium diode were used instead of the silicon device, conduction would start when there is 0.3 volt across the device.) Thus, no current flows in the circuit until after the supply has reached the 0.7-volt level. Because no current flows before the diode starts to conduct, no voltage is developed across R_L during the time period when less than $+0.7$ volt is provided by the supply. This can be derived from Equation 2-3 where the voltage

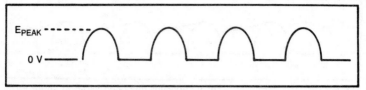

Fig. 6-5. Voltage developed across R_L.

across R_L is equal to IR_L. If I is equal to zero, the voltage developed across R_L must also be equal to zero.

Using the information that 0.7 volt must be across the diode when it conducts, the peak voltage across R_L during the conduction period must be equal to that furnished by the supply less the 0.7 volts developed across the diode. As the supply provides a 12.7-volt peak, E_{Peak} across R_L must then be 12.7 volts less the 0.7 volt or 12 volts.

When you examine the voltage supplied by a battery, it is continuous and smooth. It does not vary with time. As shown in Fig. 4-8, this voltage is fixed at a specific level, $+E$, at all times. But the voltage applied to R_L by the supply varies from 0 volt to E_{Peak} volts and then back to 0 volt during one half-cycle. It stays at 0 volt for the second half-cycle. This variation occurs 60 times each second. As a consequence, a 60-Hz tone will be heard as emanating from the radio. It is consequently desirable to eliminate this variation and get a good, steady voltage.

One method used to reduce the amplitude variations of the rectified voltage, or *ripple* is by adding a capacitor across the load. During the half-cycle when current flows and a voltage is developing across R_L, this voltage is applied to the shunting capacitor. It charges the capacitor to the 12-volt peak, $+E_{Peak}$ The capacitor holds this charge until the voltage across that component is reduced due to the cyclic variations. Now the capacitor discharges through R_L because it is wired across that load resistor. The magnitudes of the capacitor and R_L determine how long the capacitor can hold the charge before it gets fully discharged. Assuming that capacitor C and resistor R_L are quite large, the voltage across the capacitor and load resistor can be as shown by the dark lines in Fig. 6-6.

Initially, the voltage buildup across the capacitor follows the shape of the sinusoidal cycle and discharges slowly. Here, it is assumed that the capacitor is so large that it will hold the charge until the next positive half-cycle has reached a magnitude of three-fourths of E_{Peak}. It builds up to a peak during the second half-cycle and discharges until it reaches the third half-cycle, when the amplitude

161

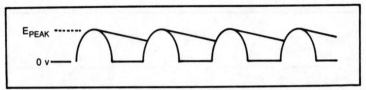

E_{PEAK} ------

0 v ——

Fig. 6-6. Voltage across R_L after a capacitor has been wired to shunt it.

of that of that curve is at three-fourths of its peak. This continues until after the ac supply has been removed from the circuit.

The first question you would probably ask, is "what advantage is there to having this capacitor in the circuit if there is still a variation in voltage during each cycle?" But note that the variation or ripple have been reduced from E_{Peak} to 0 volt to a variation of from E_{Peak} to three-fourths of E_{Peak}. Because this variation has been reduced, the audible 60 Hz has also been reduced to a considerable degree. This voltage is almost but not quite adequate to power a radio.

Ripple can be reduced further in one of two ways. One method is to increase the size of the capacitor. Now the voltage across the load will not drop as far down as three-fourths of E_{Peak}. The capacitor may be increased enough so that the voltage drops only to 99/100 of E_{Peak}, limiting the amount of 60 Hz present or ripple to a negligible level. It should be noted that if the shunting capacitor were decreased in value from its originally chosen capacity, it may be so small that it would be fully discharged long before the next cycle starts, or it may discharge to lower voltage than three-fourths of E_{Peak}.

Another way of reducing ripple is to add a low-pass RC filter after the first capacitor in the circuit and move the load to a point after that filter. This filter circuit is shown as R_F-C_F in Fig. 6-7 while C is the filter capacitor originally wired into the circuit. The C-R_F-C_F combination is known as a *pi (π) filter*. This circuit can reduce the ripple voltage to almost nothing if C and C_F are of sufficient magnitude. There is a voltage developed across R_F, but it never reaches R_L.

There are two questions you need to be concerned about. The first has to do with ripple. The second question is about the voltage. You know that a radio is usually powered by a 9-volt battery. The supply provides 12 volt peaks. Isn't 12 volts too much to power a radio?

It sure is! This is where the circuit in Fig. 6-7 comes into play. Not only can R_F in Fig. 6-7 serve as an important component in the pi filter, but it can also serve as a voltage-dropping resistor. Suppose for example, that you need 9 volts across R_L and that you have a fairly clean 12 volts across R_F. If you know the current I that you want to flow through R_L, then using Ohm's law, R_F should be made equal to 3 volts/I.

Finally, what is R_L doing in the circuit? We are talking about supplying power to a radio, not a resistor. R_L can be the resistance the radio presents to the circuit. If we know the current the radio draws, then using Ohm's law you can find that the radio presents a resistance of R_L = 9 volts/I to the supply.

VOLTAGE RATINGS, POWER RATINGS, ETC.

In an earlier chapter, I described the power ratings of resistors. Diodes and capacitor also have similar limits. A diode will accommodate only a specific reverse voltage and is capable of dissipating a specific maximum amount of power without breaking down. A capacitor is rated by the amount of voltage that can be placed across its terminals without breaking down.

Electrolytic Filter Capacitors

Lets start by noting the limits of the capacitor, C, in Fig. 6-7. The *maximum instantaneous voltage* that can be across that component is just about equal to the peak voltage of the supply. In this example, it is 1.41 times the 9 volts rms of the supply. As a safety factor, the capacitor you use should be able to accommodate more than this voltage. This is understandable when you consider that the supply voltage has a tolerance. It can vary plus or minus 10% around the center number of 120 volts. If it increases by 10% to 120 volts plus 10% of 120 volts to a maximum of 132 rms volts, the voltage across the secondary of the transformer will also increase by the 10%. In this example, this increase is to 9 volts plus

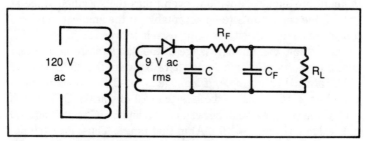

Fig. 6-7. Filter added to circuit to minimize ripple.

10% of 9 volts or a maximum of 9.9 volts rms. Therefore in this circuit, the capacitor should be able to tolerate the peak of the 9.9 volts or 9.9 × 1.41 = 14 volts, without breaking down or shorting. Another 10% safety factor is desirable, so the capacitor should be rated at a minimum of 15.4 breakdown volts. But 15.4-volt capacitors are not available. The next higher voltage standard capacitor that is available, has a 16-volt breakdown rating. so you would buy a capacitor with a 16-volt breakdown rating to fulfill your requirement.

A capacitor can also be rated by the maximum leakage current it will pass at its rated voltage and by the maximum ripple current it can handle safely.

Leakage current is measured by applying the rated voltage, E, of the capacitor to the component through a millimeter. This is shown in Fig. 6-8. Note that if a capacitor such as an electrolytic is used, it is polarized. One terminal is usually marked minus (−) and the second terminal is marked plus (+). In some cases, only one of these terminals is marked, so the second one must be of the opposite polarity. When a dc voltage is across the capacitor, the proper polarity of voltage from the supply must be applied to it. This is shown in Fig. 6-8. If the electrolytic capacitor were reversed, the leakage through the component would be extremely high. Thus, the leakage measurement for confirming that the capacitor complies with its rating is made when the proper voltage and voltage polarity is applied to its terminals.

The organization that sets the standards for the industry is the EIA. They set the standards for the maximum acceptable leakage in a capacitor. The leakage current, expressed in milliamperes, should not exceed about 0.02C + 0.3, where C is the capacity of the component in microfarads. If the leakage current exceeds this value at 25 °C, the capacitor is not acceptable. If the capacitor is used where the ambient temperature is 65 °C, the leakage at that temperature should be within seven times the specified value at 25 °C if the capacitor is to be acceptable. If the ambient temperature at which the capacitor is being used is between 25 °C and 65 °C, the acceptable maximum leakage at the temperature in question is at a proportional value somewhere between the 25 °C and the 65 °C rating. For example, if leakage is rated for a maximum of 10 mA at 25 °C, it should be less than 70 mA at 65 °C. There is a 40 °C temperature span between 25 °C and 65 °C as well as a 60 mA leakage (70 mA = 10 mA) in that range. Thus, for each degree above 25 °C, the leakage may be 60 mA/40 °C = 1 1/2 mA above the 10 mA level.

Capacitors can also tolerate a limited amount of *ripple current*, which varies in a way similar to voltage. The quantity and shape of a ripple voltage can be seen in Fig. 6-6. The peak-to-peak ripple voltage in our example is E_{Peak} minus $3E_{Peak}/4$ or $E_{Peak}/4$. The peak-to-peak ripple current is this voltage divided by the impedance of the capacitor (Equation 5-7) at 60 Hz. Rms ripple current is about one-third of the peak-to-peak in the capacitor current. Capacitor specifications should show the amount of ripple current that it can tolerate.

Unfortunately, most of the catalogs do not show the maximum ripple current or the maximum leakage current specifications of electrolytic capacitors. This information can usually be obtained directly from the manufacturer. But you can keep the ripple current quite low in a circuit simply by choosing a capacitor equal to or greater than $0.015/R_L$ farads, where R_L is the load or load resistor shunting the capacitor. As for the leakage current, you can determine what it is through use of the circuit in Fig. 6-8. Measure this current at 25 °C (77 °F) and then at 65 °C (149 °F). The latter measurement can be made by letting the capacitor sit in an ordinary temperature-controlled cooking oven for about one hour. If the second measurement is less than seven times the first one, you can assume the capacitor is in good condition. This assumes that the first reading is below the maximum current limit arrived at through use of the EIA standard equation.

To allay any apprehensions about electrolytics, however, you will find that most of them will pass these tests, so that it is not essential to perform the measurements at all times. Just buy the component from a reliable distributor with a big turnover. This will

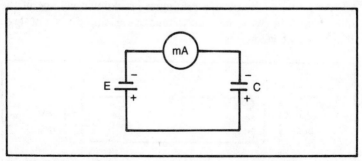

Fig. 6-8. Circuit used to check leakage current of capacitor when its rated voltage is applied to its leads. E is the rated voltage and mA is the milliameter used to measure the leakage current.

assure you that the component is of good quality and that it did not lie around on his shelves for an extremely long period of time.

Diodes

A diode can accommodate a limited reverse voltage between its anode and cathode. If a positive voltage is applied to the cathode and a negative voltage is at its anode, the diode will break down and short if that voltage is greater than a specific magnitude. Diodes are rated according to the *reverse breakdown voltage* that is necessary to cause breakdown. These ratings can be as low as a few volts or as high as several thousand volts.

Consider the circuit in Fig. 6-4. The portion of the circuit fed by the secondary winding of the transformer has been redrawn as Fig. 6-9 along with a filter capacitor. The voltage developed across the capacitor is as shown. If the capacitor is very large, this voltage remains fixed regardless of the portion of the cycle being noted. During the halfcycle when diode D does not conduct, the polarity of the voltage across the 9-volt winding of the transformer is as shown. It feeds a negative voltage to the anode of diode D with respect to its cathode. At the peak in this half-cycle, the voltage across the transformer winding is 9 × 1.41 = 12.7 volts.

Voltage across the diode is the sum of the voltage across the capacitor, 12 volts, with the voltage across the transformer, or 12.7 volts. Because of their relative polarities, these voltages add—12.7 volts + 12 volts = 24.7 volts. Thus, the rated reverse breakdown voltage of the diode must be 24.7 volts or more to accommodate this. Considering the 10% line-voltage variation, the ideal diode should have a breakdown voltage that exceeds the 24.7 volts by 10%.

As a rule of thumb, the minimum breakdown voltage of the diode should be twice the peak voltage of the supply or about three times its rms voltage.

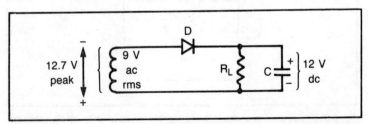

Fig. 6-9. Power supply with filter circuit. Voltages and polarities shown are during the parts of the cycle when diode D does not conduct.

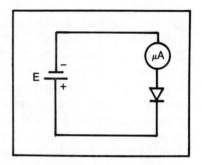

Fig. 6-10. Circuit to determine a diode's reverse leakage current.

Diodes are not perfect insulators when they are reverse biased. There is always some *leakage current* flowing through these devices. In fact, this leakage current doubles every time the diode's temperature increases by 10 °C. Good diodes exhibit low leakage-current characteristics at their rated reverse breakdown voltages. It should be low in the microampere region. The circuit in Fig. 6-10 can be used to determine the leakage of the particular diode you intend to use in a circuit. E is either the rated breakdown voltage of the diode or the actual breakdown voltage required by the circuit in which it is wired. The dc current measured through use of the microammeter is the leakage current. Make sure this current is within reasonable levels. It might show up as 0 μA on your meter. This is an ideal result.

Leakage current specifications are usually not listed in catalogs. Because of this, you should choose a component made by a reliable manufacturer. But continuous- and surge-current specifications, along with power dissipation capabilities of the diode, are just about always shown.

When a diode conducts, the *forward current* depends upon the size of the load. Assume that the load in Fig. 6-9 is R_L = 120 ohms. Because there is 12 volts across this load, the load current is 12 volts/120 ohms = 0.1A. The diode must be capable of conducting this 0.1A of current on a continuous basis.

Approximately 0.7 volt is developed across a diode while it is conducting. If it had conducted 0.1A for the full cycle, it would have to be capable of dissipating 0.7 volt × 0.1 ampere = 0.07 watts or 70 milliwatts. It actually dissipates less than half of this because it conducts for only one-half of each cycle. Even then the current is less than the 0.1A for the bulk of the half-cycle. But the diode *power dissipation rating* should be 70 milliwatts or above to include a reasonable safety factor.

When a voltage is *initially* applied to the circuit in Fig. 6-9, ca-

167

pacitor C is fully discharged. Because of this, it behaves as if it were a short circuit. The load is shorted by the capacitor at this instant. Because this portion of the circuit behaves as a short circuit, the only other components left in the circuit to limit the initial current flow are the diode and transformer winding. The only actual resistance at this time is that of the winding. It is the only resistance present to the initial or momentary *surge current*. The maximum instantaneous surge current is therefore equal to the peak voltage from the transformer winding divided by the total circuit resistance. For the circuit in Fig. 6-9, I_{Surge} = (2 volts × 1.41)/R_T, where R_T is the resistance of the transformer winding. If R_T were equal to 2 ohms, I_{Surge} = 12.7/2 = 6.35 amperes. If you had a circuit with these components, the diode you use should have a surge-current rating of more than 6.35 amperes.

DIODES IN SERIES

In our example, you need a diode with at least a 25-volt breakdown capability. There is no problem getting a diode with this low breakdown voltage. Diodes are readily available with specified breakdown voltages of 100, 200, 400, 800, 1000, and more volts. Although higher voltage diodes are available, they are not commonly used.

Suppose you need a diode with a 2000-volt breakdown characteristic and you only have several 1000-volt devices in your possession. You would immediately assume that you should connect two 1000-volt diodes in series, for a total breakdown voltage of 2000 volts. This is definitely a logical conclusion. You assume that if you do this, the applied 2000 volts should divide evenly between the two components so that there is 1000 volts across each device. But this is true only in an ideal situation.

The leakage currents of diodes with the same breakdown voltage are all different. Because of this, the applied voltage will not divide equally between the two diodes. Less than 1000 volts will be across the device with the higher leakage and the balance of the 2000 volts, or more than 1000 volts, will be across the device with the lower leakage. The latter diode may break down because of this excessively high reverse voltage.

A logical alternative is to add a third diode to this series arrangement so that in the ideal case, 1/3 of 2000 volts or 667 volts is across each device. This gives each diode a safety factor when used in the circuit.

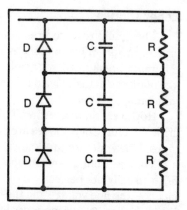

Fig. 6-11. Circuit to add accuracy for an equal division of applied voltage between the series-connected diodes.

A rule of thumb that can usually be used to determine the breakdown voltages is to multiply the circuit voltage by 5/4 or by 1.25. Next, get diodes with equal breakdown voltages and add to at least this figure. Thus, if the supply is 2000 volts, 5/4 times 2000 = 2500 volts. So if you want to use two diodes, the breakdown voltage of each device must be equal to or more than 2500/2 = 1250 volts. Should you want to use three diodes, the breakdown voltage of each device must be at least 2500/3 = 833 volts. If you have lower voltage devices on hand, divide 2500 by the breakdown voltage of the diodes you have. The quotient is the number of these diodes you should use in the circuit. For example, if you have 500-volt diodes, you would have to connect 2500/500 = 5 devices in series, to satisfy the situation.

To further assure that the voltage will divide more-or-less evenly between the devices, a capacitor and resistor can be connected across each diode. This is shown in Fig. 6-11. Each capacitor is usually about 0.005 μF. It must be able to accommodate a voltage equal to the rated breakdown voltage of the diode it is shunting. Each resistor should be equal to or less than $500V_B/I_R$ (max) where V_B is the breakdown voltage of each diode and I_R (max) is the maximum reverse-diode leakage current specified by the manufacturer of the device. If no specification is available, assume I_R (max) = 1 mA or 0.001 ampere.

SPECIAL DIODES

Diodes are also specially designed and/or selected to perform tasks other than power rectifiers. Among their many applications, they serve as voltage regulators, radio signal detectors, light-sensitive devices, capacitors, and so on.

Zener Diodes

The actual reverse breakdown voltage of a rectifier is not a critical factor. When a diode is used in this application, only the minimum breakdown voltage is important. The breakdown voltage of diodes actually used in a circuit can be any value above this minimum.

However, there are circuits in which the reverse voltage is a critical factor. Some diodes are built to break down at specific voltages. These device are known as *zener diodes.*

Voltage from a power supply such as the one shown in Fig. 6-7 is usually not perfectly fixed. This voltage may vary with two factors—the line voltage and the resistance that is used as the load. When the line voltage increases by 10%, from 120 volts to 132 volts, the voltage across the secondary winding will increase by the same 10% from 9 volts to 9.9 volts. Should the line voltage drop by 10% to 108 volts from 120 volts, the 9 volts will drop by the same 10% to 8.1 volts. As a resu¹⁺, the rectified and filtered dc voltage will increase and decrease with the line-voltage fluctuations.

Next consider what the load voltage is when the magnitude of the load changes. In Fig. 6-7, R_F and R_L form a voltage divider. If R_L is small, less of the voltage that is across C is across R_L than if R_L were large. If R_L is large, a bigger portion of the supply voltage is across the load than if R_L were small. The voltage across R_L varies with the size of this load resistance.

The variations of the supply voltage and load may have no effect on the equipment or circuit being powered. But there are circuits that require a fixed or stable dc supply. The zener diode can be placed across the varying supply to regulate the voltage fed from it to these circuits. Voltage across the zener diode remains relatively constant despite line voltage and load variations.

The voltage across any load connected in parallel with this diode is equal to the breakdown voltage of the diode. A circuit with a regulated output voltage is shown in Fig. 6-12. Diode D_1 is the rectifier supplying an unregulated 12 volts across capacitor C, while the regulated voltage across the load resistor, R_L, is the same as the fixed reverse breakdown voltage of zener diode D_2 connected across that resistor. R_S is an isolation resistor placed between the unregulated voltage across C and the regulated voltage across D_2. I assume that 9 volts is required across the load so a diode D_2 is chosen which breaks down when the voltage at its cathode is 9 volts more positive than the voltage at the anode.

In our first example, assume that R_L is fixed at 1000 ohms and

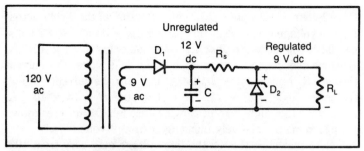

Fig. 6-12. Power supply with a regulated voltage at its output.

that it does not vary from that resistance. Because 9 volts is across the resistor, it conducts 9 volts/1000 ohms = 9×10^{-3} A. The circuit is usually designed for the zener diode to conduct a current equal to at least one-tenth the current conducted by the load. So if the load conducts 9×10^{-3} A, the diode should conduct 0.9×10^{-3} A. The total current flowing in the parallel-connected components, R_L and D_2, is the sum of these two currents of 9.9×10^{-3} A. This total current must also flow from the supply through R_S to the load resistor and zener diode.

Voltage at the righthand edge of diode D_1 and across C is 12 volts dc. But when the line voltage drops by 10%, the dc voltage can drop by 10% or by 1.2 volts. Now the voltage across the capacitor would be 12 volts less the 1.2 volts or 10.8 volts. Resistor R_S must be capable of passing the 9.9×10^{-3} A even when the dc across the capacitor is 10.8 volts. The lowest voltage that can be across R_S is when the supply is at a minimum. At this time, the voltage across R_S is 10.8 volts − 9 volts = 1.8 volts. Dividing this voltage by the current that must flow through R_S, you will find that this resistor must be equal to or less than 1.8 volts/9.9×10^{-3} amperes = 182 ohms.

When the supply voltage is at its normal or rated level, 12 volts is across C. Now the voltage across R_S is 12 volts − 9 volts = 3 volts, so that the current it passes under these conditions is 3 volts/182 ohms = 16×10^{-3} A. As a consequence, D_2 must now conduct 16×10^{-3} A − 9×10^{-3} A = 5×10^{-3} A. Even though R_S conducts more than the 9.9×10^{-3} A current than it did when the line voltage was low, the voltage across the zener diode remains fixed at 9 volts. Because it is fixed at 9 volts, it keeps the voltage across R_L constant at this same magnitude.

The zener current increases further when the supply voltage is 10% above it center value. Now the unregulated voltage across

C is 12 volts + 1.2 volts = 13.2 volts. Because of the 9 volts across D_2, the voltage across R_S is now 13.2 volts − 9 volts = 4.2 volts, so that it conducts a total of 4.2 volts/182 ohms = 23×10^{-3} A. Because the voltage across R_L is fixed at 9 volts, the current through R_L remains at 9×10^{-3} A. But now the current through the zener diode is 23×10^{-3} A − 9×10^{-3} A = 14×10^{-3} A. Even with this high current flowing through D_2, its breakdown voltage remains relatively unchanged from 9 volts.

Note that even though the line voltage may vary from 10% below its center value of 120 volts ac to 10% above it, the voltage across the zener is fixed. I assumed that sufficient zener current flows through the diode when the supply voltage is at a low level. I must also assume that the zener diode is capable of dissipating sufficient power when the line voltage is at its maximum. Because the zener current is now at its 14×10^{-3} maximum and there is 9 volts across the diode, as always, it will dissipate a maximum of 14×10^{-3} A $\times$ 9 volts = 0.126 watts. It must be capable of dissipating at least this amount of power.

As a second example, suppose the line voltage remains steady at the 120 volt ac level so that there is always 12 volts across capacitor C. You want the voltage to remain fixed at 9 volts regardless of whether R_L is 2000 ohms or 200 ohms.

First, you must find the maximum resistance that R_S may be. This can be determined by initially considering the condition when R_L is at its minimum. In our example, this is R_L = 200 ohms. The current through the 200-ohm resistor is 9 volts/200 ohms = 45×10^{-3} A. Since the minimum zener current should be about one-tenth of this or 4.5×10^{-3} A, the total current that flows through the lead and zener is 45×10^{-3} A + 4.5×10^{-3} A = 49.5×10^{-3} A. This current must also flow through R_S. Because the voltage across R_S is 12 volts − 9 volts = 3 volts, the maximum resistance of R_S is 3 volts/49.5×10^{-3} A = 61 ohms.

When R_L is at its maximum or 2000 ohms, the current through that resistor is 9 volts/2000 ohms = 4.5×10^{-3} A. The balance of the 49.5×10^{-3} A flowing through R_S, or 49.5×10^{-3} A − 4.5×10^{-3} A = 45×10^{-3} A, must then flow through the zener. The power dissipated by this zener is now (45×10^{-3} A) (9 volts) = 0.405 watts. You must use a zener with at least this rating.

You should not forget that R_S also dissipates power. The current it conducts is 49.5×10^{-3} A, while the voltage across it is

3 volts. The product of the two or 0.149 watts is dissipated by this 61-ohm resistor.

Varactor Diodes

The diode structure shown in Fig. 6-2, can be split into three sections. In the one constructed of n-type material, you have floating electrons. In the second one that is formed using a p-type semiconductor slab, there are floating holes. Between these two sections you have a depletion region. Once the electrons and holes have settled in this region, they stop floating. But the remaining electrons and holes keep floating around in this region of the n-type and p-type slabs.

The n-type and p-type materials are effectively conductors because the electrons and holes are not fixed in place. They can float around inside their respective slabs. The areas near the junction of the semiconductors do not have freely floating electrons or holes. Consequently, this depletion region behaves as if it were an insulator. You then have an insulator sandwiched between two conductors. This is very much like the situation depicted in Fig. 5-11 showing the basic structure of a capacitor. Because of this, the diode also behaves as a type of capacitor. It is essentially a variable capacitor. Capacity varies with the voltage applied between the anode and cathode.

When there is zero voltage applied simultaneously to the n-type and p-type materials, the width of the depletion region is fixed as shown in Fig. 6-2. Because of this, the capacity is fixed at a specific value. Let us refer to this capacity as C.

Should the n-slab (or cathode of the diode) be made positive with respect to the p-slab (or anode), the diode is reverse biased. The depletion region gets expanded because of the presence of these voltages. Because the insulating depletion region is wider, the capacity of the diode is reduced. Thus, the capacity of a diode decreases as the reverse voltage is increased. Capacity is now less than C.

A silicon diode does not conduct any current until after the forward voltage exceeds about $+0.2$ or $+0.3$ volts. The insulating depletion region exists while this small positive voltage is applied to the anode with respect to the cathode. But the width of the depletion region is reduced as the applied positive voltage keeps increasing to the point just before it starts to conduct. Because the width of the depletion region is reduced, the conducting areas of the n-type and p-type slabs get closer to each other. The capacity of the diode is thereby increased. But the forward voltage is limited to magnitudes which are small enough so as not to cause the diode to start conducting. Under these conditions, the capacity of the diode is more than C.

A parallel resonant circuit using two varactor diodes is shown in Fig. 6-13. If $+E$ and $-E$ are equal, the voltage at the moving arrow or wiper (and anodes of the diodes) is zero when it is at the center of the fixed resistor or control. Zero volts is also at the common or ground return terminals of the batteries supplying the $+E$ and $-E$ voltages. It is likewise zero at the cathodes of varactors D_1 and D_2. The common ground is connected directly to the cathode of D_2, but is connected through the low resistance of inductor L to the cathode of D_1, putting that terminal at 0 volt. Each diode has a specific capacity when the voltage across that device is zero. Since the capacity of D_1 is in series with the capacity of D_2, their total capacity (or the capacity across L) can be calculated in the same manner used to calculate the capacitance of any two capacitors connected in series. It is determined using Equation 3-2 where the two C's are substituted for the R's in the equation. The C determined from this equation is in parallel with L. The resonant frequency of the combination is determined with reasonable accuracy through use of Equation 5-12.

When the wiper arm of the control is moved above center to near the top of the fixed resistor, positive voltage is placed on the anodes of the varactors with respect to their cathodes. This would increase their capacities over what they were when the moving arm was at the center. The capacities would be reduced if the wiper

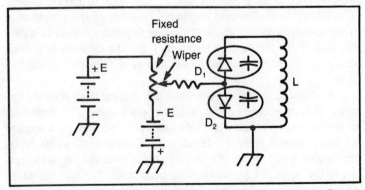

Fig. 6-13. Parallel resonant circuit. Tank circuit is tuned by varactors. Capacities of diodes are adjusted by voltage applied to circuit.

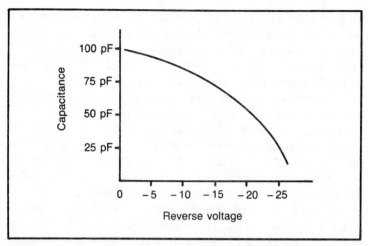

Fig. 6-14. Graph showing how the capacity of a varactor may vary with the voltage applied to its terminals.

arm were move to below center of the control, for now the diodes would be reverse biased. In either case, the resulting resonant frequency is determined from the capacity of the series-connected varactors in conjunction with L. The desired resonant frequency is adjusted by applying different voltages to the diode by placing the moving arm into different settings.

Information as to the capacities of a varactor diode (also referred to as a *voltage-variable capacitor* or a *varicap*) when different voltages are applied is usually supplied in data sheets. There are graphs provided by many of the manufacturers that show how the capacity varies with the applied voltage. One such curve is shown in Fig. 6-14.

Photodiodes

If a diode is reverse biased, a leakage current will flow through it. This is because the diode, like any other device, is not a perfect component. Should the diode be placed in an oven and the temperature increased, more leakage current will flow than at low temperatures despite the fact that there is a fixed reverse voltage across its terminals. As noted earlier, leakage current doubles every time the reverse voltage is increased by 10 °C.

Why does leakage current increase with temperature? Because heat is a form of energy, as the diode's temperature is increased, there is more energy applied to it. When applied to electrons at the junction of the semiconductor slabs in the diode, this increased energy, activates the floating electrons at high temperatures more than at normal room temperature. Because of this, these energized electrons have a greater tendency to flow through the diode than they did before, and a large current is generated.

But light is also a form of energy. If light hits the junction of a reverse-biased diode, its leakage current increases over what it is when the diode is in a dark environment. Most diodes are enclosed in an opaque container that will not let light pass to the junction. Consequently, these devices are not affected by the presence of light. Special devices are placed in a clear plastic holder to allow the light to hit this junction. These devices are known as *photodiodes*.

The current passed by a photodiode varies with the amount of light hitting the junction. If that light is bright, more light energy would hit the junction to allow more reverse current to flow than when the light is dim.

There are a number of worthwhile applications using this phenomenon. One of these is the light-activated switch. A basic application is shown in Fig. 6-15.

Battery E reverse biases photodiode D through the relay coil. The symbol of the photodiode is identical with that of an ordinary junction diode, except that there are two arrows to show that light can hit the junction. In the dark, very little current flows through D and the relay coil.

A relay is a device constructed from a coil and a moving arm. The arm remains up in the position shown when there is no force to pull it toward the coil. When the arm is up, it makes contact with a switch terminal, completing a circuit. In the figure, it is shown as completing a circuit with a 120-volt ac supply and a bulb, so that the bulb lights when the arm is in the position shown.

When sufficient current flows through the coil, a magnetic field is formed with enough strength to attract the moving arm to the coil. When this is accomplished, the 120-volt ac bulb circuit is broken and the light is extinguished. This is where the diode comes into play.

Assume the diode is placed in an area where it can sense whether it is light or dark outside of a house. In the circuit, it can be used to actuate the relay when it is light and deactivate it when it is dark. During the night when it is dark, very little current flows through the diode. What current does flow is insufficient to actuate the relay. The switching circuit remains closed and the light stays lit.

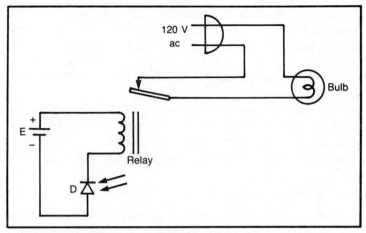

Fig. 6-15 Basic circuit of light-activated switch.

When daylight arrives, the current in the diode and relay is increased sufficiently to trip the relay. The magnetic field is strong enough so that the moving arm is pulled down and the switch contacts open. Now the light goes off.

All this assumes that the photodiode is not located near the bulb so that it is not activated by the light the bulb sends out.

Photovoltaic Diodes

A voltage is developed across a specially designed diode when its junction is illuminated. There is about 0.6 volt present at the leads of this *photovoltaic diode*. When several of these diodes are connected in series, this voltage is multiplied by the number of diodes being used. Should the series-connected diodes be placed across a rechargeable battery, it will be charged by these devices.

Light-Emitting Diodes

Just as heat can be applied to a diode to increase its reverse or leakage current, heat can be created inside the diode when forward current or large amounts of reverse current flow through it. Heat is one form of energy that may be present due to this current. A second form of energy generated by this means is light. This is the basis of the *light-emitting diode* or LED.

LEDs are now used as pilot lights in different pieces of equipment rather than small bulbs with ordinary filament or wire conductors. This is because LEDs require less current for them to light than conventional bulbs. Consequently, LEDs draw less power from a supply. The circuit feeding current to the LED is designed to limit the maximum current to the rated amount required by the diode to light.

The most common color is red, but they are available in many different colors.

The schematic representation of an LED is the same as the diode in Fig. 6-15, except that the arrows are at the opposite ends of the lines and point in the reverse direction from that shown here. This is to indicate that light is emanating from the diode.

Point Contact Diodes

Not all diodes are constructed from two semiconductors. Point contact diodes consist of a metal point that makes contact with a semiconductor material. It conducts or is forward biased when the metal is positive with respect to the semiconductor. Now the metal is the anode and the semiconductor is the cathode. The voltage across this junction is very low when it is forward biased—usually around one-fourth of a volt.

The amount of current the diode conducts is fairly limited when it is reverse biased. Also, point-contact diodes tend to suffer from greater reverse leakage currents compared to junction diodes.

So if it is not such a great component, why do we bother with it? Actually, they have a lot of good applications that cannot be filled by junction diodes. Diodes are needed for use at extremely high frequencies up to and including the GHz region. Point contact diodes serve quite well in these special high-frequency applications.

Microwave diodes are also known as *Schottky-Barrier* diodes. New point contact diodes have been designed to further improve performance at these frequencies over what is possible with the older types. These are known as *hot-carrier* diodes.

Tunnel Diodes

In electronic circuits, you may commonly find one more group of junction diodes—the *tunnel diodes*. This diode is also used at relatively high frequencies. It is quite versatile because it can be placed in circuits to perform either as a switch, an oscillator or signal generator, or an amplifier. The various symbols used by the industry for this diode are shown in Fig. 6-16. Its particular function in a circuit is usually indicated by the manufacturer of the equipment using that circuit.

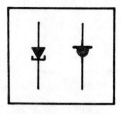

Fig. 6-16. Commonly used schematic symbols for a tunnel diode.

Unlike all other diodes, there are several independent regions in the graph of the tunnel diode's operating characteristics. In one, the forward current flowing through the device increases as the applied voltage increases. However, is true of just about every diode. But in another region, the current flowing decreases as the applied voltage increases. It is this latter characteristic which makes the tunnel diode such a useful and versatile device.

The tunnel diode is used relatively infrequently now because it has been replaced by more economical and efficient integrated circuits. It is discussed here so that you will know what it is if you should ever see it in a circuit.

Chapter 7

Power Supplies

Circuits using diodes as rectifiers and capacitors as filters in power supplies, were discussed in Chapter 6. Because only one-half of the sine wave was used to provide the power for conversion to dc from the ac at the input, this arrangement is referred to as a *half-wave* power supply. But there are supplies with better characteristics that utilize both halves of the sine wave.

FULL-WAVE SUPPLY

In Fig. 6-6, the capacitor is shown as discharging to about three-fourths of E_{Peak}. If the circuit were modified to supply another half-cycle of voltage between each of the two half-cycles shown in the drawing, the voltage across the load resistor and filter capacitor would not drop to as low as three-fourths of E_{Peak}. It would remain above this voltage even if the same filter capacitor were used across the load here as was used in the original half-wave circuit.

The output with two half-cycles is shown in Fig. 7-1. Here, the voltage across the filter capacitor would drop from the peak to somewhat above 7/8 of E_{Peak} before it is recharged by the next half-cycle of voltage. A circuit capable of supplying the waveshape with two half-cycles is shown in Fig. 7-2A.

The full-wave power supply is essentially two half-wave circuits connected across the same load. In the drawing, ac at 120 volts rms is applied to the primary winding of the power transformer as was done in the half-wave circuit in Fig. 6-4. There, 9 volts rms

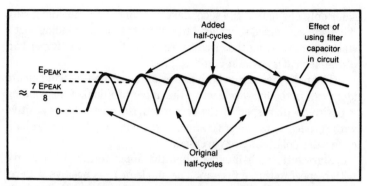

Fig. 7-1. End result of adding a half-cycle of voltage between the two half-cycles of voltage in a half-wave power supply.

is present only across the entire secondary winding. In Fig. 7-2, 9 volts rms is across only one-half of the winding. An identical 9 volts is also across the second half of this winding, so that a total of 18 volts rms is across the entire secondary. This transformer is specified as a 120-volt to 18-volt component, with a center-tap on the secondary winding. In describing the circuit, you need con-

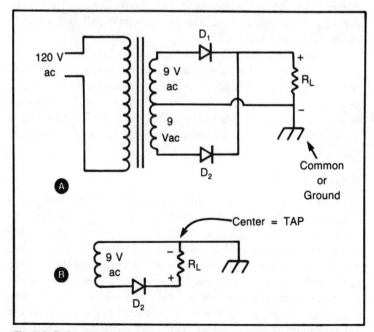

Fig. 7-2. Full-wave power supply. A. Complete circuit. B. Circuit showing how current flows through D_2 to R_L.

sider only one-half of the secondary winding at a time, along with the diode connected to the lead from that half of the winding. Let's start by considering the circuit formed by D_1 with the upper half of the secondary winding and R_L.

The polarities across both halves of the secondary winding are identical. In the half-cycle when the upper terminal of the winding is positive with respect to the center-tap, the center-tap is positive with respect to the lower terminal of the winding. In the next half-cycle, the polarities are reversed.

Start with the half-cycle when the upper terminal is positive with respect to the center-tap. The anode of D_1 is positive with respect to its cathode. Consequently, current flows through this diode and through R_L. The half-cycles of voltage developed across the load resistor due to this current are shown in Fig. 7-3A.

The circuit involving the lower half of the winding and D_2, is redrawn for clarity in Fig. 7-2B. When the center-tap of the secondary winding is positive with respect to the bottom lead, diode D_2 does not conduct. Only D_1 conducts. Diode D_2 conducts only on the alternate half-cycles when the polarity of the leads are reversed. The half-cycle of voltage developed across R_L because of current flowing through D_2 is shown in Fig. 7-3B. During these half-cycles, D_1 does not conduct. The two groups of half-cycles shown in the figure are 180° apart.

Current from both diodes flows through R_L. One diode conducts for one-half of the cycle while the other conducts for the second half of the cycle. Voltage is developed across R_L because of these two current sources. The total voltage across R_L is the sum of the two voltages shown in Figs. 7-3A and 7-3B. These voltages are added because they are both across R_L. The sum is shown in Fig. 7-3C.

Filters

The effect of placing a filter capacitor across the load to reduce ripple at the output of a full-wave supply is shown in Fig. 7-1. Ripple is certainly much lower here than it was with the identical capacitor across the load in the half-wave circuit. In this circuit, ripple can be reduced to reasonable levels by making the product of R_L and C (a shunting capacitor) equal to 0.1, when R_L is expressed in ohms and C in farads. This will not only reduce the ripple so that it is acceptable in most applications, but will also provide the maximum dc output voltage this circuit is capable of producing.

For half-wave circuits, the product of R_L and C should be

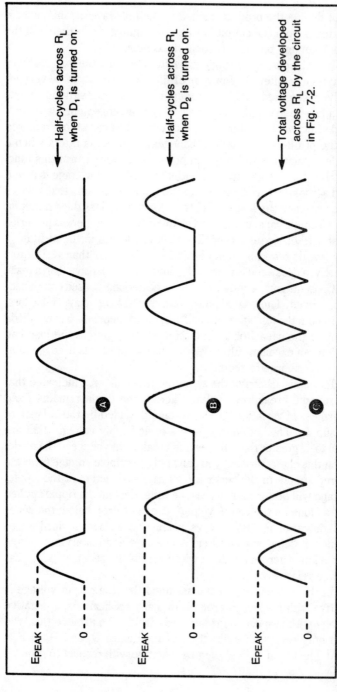

Half-cycles across R_L when D_1 is turned on.

Half-cycles across R_L when D_2 is turned on.

Total voltage developed across R_L by the circuit in Fig. 7-2.

E_{PEAK}

0

E_{PEAK}

0

E_{PEAK}

0

Ⓐ

Ⓑ

Ⓒ

Fig. 7-3. Individual voltages developed across R_L by pulsating dc currents of circuit in Fig. 7-2A. The sum of these voltages is also shown.

about double the product needed in the full-wave arrangement if the ripple from the circuit is to be reasonable. Only then will the output voltage be near its ideal maximum.

It is obvious that a single capacitor shunting the load is a very basic type of filter. The pi type of filter was shown to be very effective when used in half-wave supplies. A similar situation exists when full-wave supplies are used. Pi type arrangements can be applied effectively as filters in both circuits. The capacitors required for the pi filters are smaller than those needed as filters when a single capacitor is the only component used to perform this function. But there is a drawback with the pi filter. A voltage is developed across the series resistor in the filter circuit, and that voltage never reaches the load. It is lost power. This loss does not exist when a single capacitor is used as the filter. Here there is no series resistor to dissipate power. To minimize losses in the pi filter, a very low dc resistance choke can be used rather than R_S. Unfortunately, a choke coil is more bulky and more expensive than a resistor. Consequently, a resistor is the most frequently used component in this circuit. The wasted power is accepted as an unavoidable loss.

In both the full-wave and half-wave arrangements, a zener diode shunting the capacitor filter circuit or wired across the load can serve as an excellent filter. When this component is used, smaller filter capacitors are required.

In the discussion of the zener diode regulator, I indicated that the reverse breakdown voltage across the diode remains fixed regardless of variations in the voltage provided by the balance of the supply. This, of course, refers to dc supply voltages that are equal to or greater than the reverse breakdown voltage of the diode. Using this characteristic, you can apply the ripple voltage from the filtered supply to the zener for a further reduction of the ripple. Voltage will be constant across the zener despite the ripple applied to it as long as the ripple voltage does not drop below the zener breakdown voltage. Because a constant voltage is maintained across a zener, it behaves as a filter. A constant and unvarying voltage is developed across any load wired across the zener, making that voltage relatively ripple free.

In this discussion, I have assumed that the output voltage is positive with respect to ground. In many applications, a negative voltage with respect to ground is desirable. To achieve this, you need only reverse the direction of the diodes D_1 and D_2 in Fig. 7-2A. The top of R_L will then be negative with respect to ground.

In half-wave or regulator circuits, all diodes must likewise be reversed to achieve this goal.

BRIDGE RECTIFIERS

Although the full-wave power supply system described here is quite good, efficiency is not great. This can be improved through use of an alternate circuit. The bridge circuit in Fig. 7-4 is used quite frequently to fulfill this goal. Here, four diodes are used rather than only two as shown in Fig. 7-2. But the circuit has the important economy advantage over the standard full-wave arrangement in that a less expensive transformer can be used here to drive the circuit. Even the rectifiers are less expensive because the reverse breakdown voltage due to ac peaks at the transformer is split between two of the diodes. Since each diode need only accommodate about half the reverse voltage as in the previous circuit, the diodes are also somewhat less expensive. But you can't just divide the breakdown voltage by two to determine the rating of each of these diodes. You must then multiply the breakdown voltage by 5/4 to take into account the variability in diode breakdown characteristics as noted previously.

The operation of the circuit is quite simple. When the polarity of the ac voltage at the 9-volt winding is such that the upper terminal of the winding is positive with respect to its lower terminal, current flows through D_1, R_L and D_4. This is because only the anodes of these diodes are more positive than their cathodes. A half-cycle of voltage is developed across R_L due to the current flowing through it. No current flows through D_2 and D_3 because their anodes are negative with respect to their cathodes. If a filter

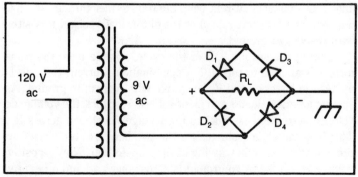

Fig. 7-4. Power supply using bridge rectifier circuit.

capacitor is across R_L, the total peak reverse voltage across the two diodes when off is equal to twice the peak of the applied ac. This voltage is divided between diodes D_2 and D_3.

On the alternate half of the cycle, diodes D_2 and D_3 conduct current through R_L, while diodes D_1 and D_4 do not conduct. A half-cycle of voltage is also developed across R_L because of this current. The sum of the two half-cycles across R_L is as shown in the bottom diagram in Fig. 7-3.

Ripple is present here across R_L just as it was at the output of the standard full-wave circuit. As before, a capacitor shunting R_L will help diminish any undesired ripple. To accomplish this, the product of R_L and C should be greater than 0.1, just as in the circuit in Fig. 7-2.

In Fig. 7-4, the negative terminal of R_L is shown to be at ground so that the output is positive with respect to ground. To get a negative voltage with respect to ground, either reverse all of the diodes or just connect the ground lead to the opposite or + end of the load resistor. The remaining end of the resistor will now have a negative voltage with respect to ground.

POSITIVE/NEGATIVE SUPPLIES

Different pieces of equipment require that positive and negative voltages be supplied at the same time from one voltage source. There is an attempt in many cases to keep the two voltages of opposite polarities equal. Two unregulated circuits that are designed to accomplish this are shown in Fig. 7-5.

The arrangement in Fig. 7-5A consists of two identical half-wave circuits. D_1 is the same here as it was in the circuit in Fig. 6-4. The only difference is that here a filter capacitor, C_1, has been added. D_1 conducts current from the transformer through R_{L1} during one-half of the ac cycle. The top of this load resistor is positive with respect to ground.

During the second half of the cycle, voltage from the transformer is applied to the D_2/R_{L2} combination. Current through D_2 makes the lower end of R_{L2} negative with respect to ground. Assuming identical diodes and identical load resistors, the voltage developed across R_{L1} is equal to the voltage developed across R_{L2}. This is because the voltages for the two halves of the circuit are provided by the same winding of the transformer. Now a positive voltage with respect to ground is across one load and a negative voltage with respect to ground is across the second load.

A similar situation exists in the circuit in Fig. 7-5B, except that

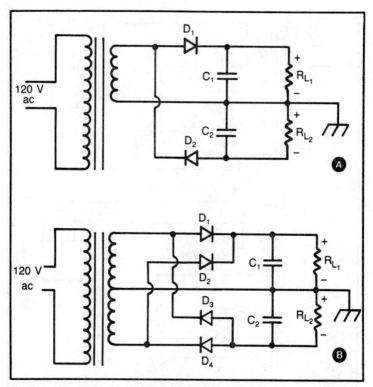

Fig. 7-5. Two circuits providing just about identical positive and negative voltages with respect to ground. A. Half-wave circuit. B. Full-wave circuit.

here two full-wave power supplies are used. D_1, D_2 and R_{L1} form the circuit shown in Fig. 7-2, while D_3 and D_4, along with R_{L2}, form an identical full-wave circuit with the diodes wired in opposite directions. In this case, the voltage across R_{L2} is negative with respect to ground, but it is close to being identical in magnitude to the voltage across R_{L1}. Voltage across R_{L1} is, of course, positive with respect to ground.

Two bridge circuits can be used in a similar fashion to provide identical positive and negative voltages. For this application, transformers are frequently provided with two identical windings—each one to feed bridge circuit. This is shown in Fig. 7-6A. Bridge circuits can also be used with a transformer which has only one secondary winding. This is shown in Fig. 7-6B.

The bridge circuit shown in Fig. 7-6C may also be used to supply positive and negative voltages to two loads at the same time. If identical currents are required through R_{L1} and R_{L2}, they may

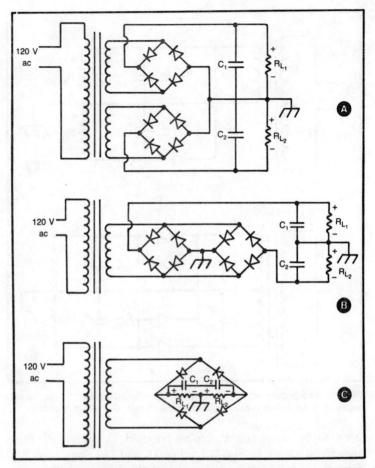

Fig. 7-6. Two bridges supply positive and negative output voltages. A. Each bridge uses a separate but identical transformer winding. B. Only one transformer winding is used for both bridges. C. Only one bridge is used here. Positive and negative voltages are available because the load is split into two equal parts.

be wired in series to the bridge. A ground may then be placed between the two loads. The sum of the voltages developed across the two resistors is equal to the voltage developed across the single load shown in Fig. 7-4 if the transformers and loads across the two circuits are identical. But here, the two individual voltages are significant. Because the same current flows in both resistors, the voltages developed across the resistors depend upon the relative magnitudes of the components. Individual voltages are determined by using Ohm's law.

188

Whatever circuit is used, the output can be regulated using any of the arrangements described in the previous chapter or discussed below.

VOLTAGE DOUBLERS, TRIPLERS, ETC.

Unregulated voltages available from half-wave or full-wave supplies may be too low to do the job for a particular circuit. The dc derived from a power supply with one transformer may not be sufficient for powering a specific circuit and keep it performing properly. This is frequently the case where extremely high dc voltages are needed.

To correct this, the first thought that comes to mind is to use a transformer with a high voltage at its secondary winding. This can usually be done, but the price of the transformer can be exhorbitant. The high price is because of the good insulation required to avoid electrical arcing and breakdown in the transformer due to the high voltages across and between the windings. Hence, a lower voltage component may be preferable along with some sort of voltage multiplier circuit at its output.

In all the examples, I showed a transformer being used between the power lines and the dc supply. When high dc voltages are required, the transformer may not be needed in the circuit. Voltage may in some instances be taken directly from the power line. An example applying this method using a half-wave supply is shown in Fig. 7-7.

One-half of each cycle is passed through resistor R_S and diode D to the load, R_L, and filter capacitor, C. The peak voltage present during each half-cycle is 120 volts times 1.414 or 170 volts. After passing through the rectifier, the half-cycles are filtered by capacitor C so the dc across the capacitor and R_L is just slightly below

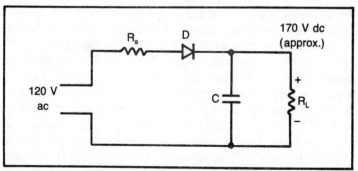

Fig. 7-7. Half-wave dc power supply using voltage directly from the power lines.

the peak voltage. Here, I will assume it to be equal to the peak voltage. The percent error due to this assumption is slight because the output voltage is considerably more than the usual voltage drops in R_S and D.

R_S is a very small resistor. Its function is to limit the surge current through diode D to a safe value. If, for example, the diode is permitted a surge current of 15 amperes when the circuit is first turned on, R_S must be equal to somewhat more than (120 volts) (1.414)/15 amperes = 170/15 = 11.3 ohms. A 12-ohm resistor can be used here. Because R_S is very small, it should have little or no effect on the load current flowing through the circuit after the dc voltage across C and R_L has stabilized to a constant 170-volt level. All other considerations apply the same way here just as in Fig. 6-4, after a capacitor has been added across R_L in that circuit.

You must take certain precautions when handling this type of circuit. Because it is connected directly to the power line, do not connect any leads to any ground point. Also, do not connect any wires from the supply to a radiator, water pipe, oven, refrigerator, and so on. Always be very careful when working with this circuit. If you touch the wrong lead and ground yourself at the same time by standing on the earth or by touching any grounded appliance, it can be fatal. So if you do not use a transformer, take these extra-special precautions. Remember that you are no longer working with a mere 9 or 12 volts. Also, it may even be lethal if you have one hand on one lead from the power line and the other hand on the second lead, and if you touch one end of R_L or C with one hand and touch the second end of either component with your other hand. In both situations, current may flow through the heart and do irreparable damage or kill! But with proper precautions, there is no problem in taking voltages directly from the power lines.

If the line or transformer secondary voltage is too low, one of the doubler circuits shown in Fig. 7-8 may be used. Voltage can be supplied to the circuits directly from the power lines, as shown, or from the secondary winding of a power transformer.

In Fig. 7-8A, one diode conducts at a time. Diode D_1 conducts during the half-cycle when the upper terminal of the power line is positive with respect to its lower terminal. The 170 volts dc is developed across C_1. During the alternate half-cycle, D_2 conducts so that 170 volts is now developed across C_2. Voltages across the two capacitors are added. Because of the relative polarities of these voltages, the total voltage across the two capacitors is 170 volts + 170 volts or 340 volts dc. This is applied to the load, R_L.

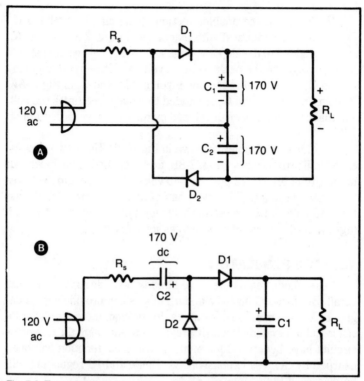

Fig. 7-8. Two voltage-doubler circuits. Both use half-wave rectifier arrangements.

Similar logic can be used to analyze the circuit in Fig. 7-8B. During the half-cycle when the lower terminal of the ac line is positive with respect to its upper terminal, diode D_2 conducts. Capacitor C_2 is thereby charged to 170 volts with the polarity shown. During the alternate half-cycle, the upper terminal of the supply is positive with respect to the lower terminal. At the peak of this half-cycle, the 170 volts at the supply adds to the dc voltage previously developed across C_2. The two voltages add to an instantaneous peak of 170 volts + 170 volts = 340 volts. This is applied to the circuit of D_1, C_1, and R_L. Now D_1 conducts. The sum of the 170 volt dc with cyclic peaks of 170 volts is thereby developed across C_1 and R_L. C_1 filters this voltage so that a steady 340 volts is across the capacitor and R_L.

Using combinations of the two arrangements in Fig. 7-8, the input voltage can supply dc that is triple, quadruple, and so on, of the voltage developed by the simple half-wave or full-wave circuit. Examples of this are shown in Fig. 7-9.

In Fig. 7-9A, the doubler section of the circuit involving D_1, D_2, C_1, and C_2 is identical with that shown in Fig. 7-8B. The D_3/C_3 circuit is added to develop a voltage across C_3. This voltage should be about equal to the peak of the ac voltage. The circuit involving D_3 and C_3 is identical to the one involving D_2 and C_2 in Fig. 7-8A. Voltages across C_1 and C_3 are added to equal three times the voltage that could have been developed by the half-wave circuit in Fig. 7-7.

A voltage quadrupler is shown in Fig. 7-9B. Here, two circuits, each similar to the one in Fig. 7-8B, are combined. The upper half of the circuit provides a positive 340 volts with respect to the lower lead from the supply. The lower half of the circuit provides a negative 340 volts with respect to this same lower lead. The two voltages are added and applied across load resistor R_L.

VOLTAGE REGULATORS

Zener-diode regulators have already been described in some detail. But forward-biased standard diodes can provide a good degree of regulation in some instances. In addition, there are also transistor and integrated-circuit (IC) regulators. Although the transistor circuits have been used for many years, their function has been usurped by the less expensive ICs. Diode and IC regulators will be discussed here.

Diode Regulators

There is a constant voltage across the ideal zener diode when it is reverse biased. You will find that this voltage remains relatively constant despite the quantity of current flowing through the zener diode when its reverse breakdown voltage is greater than 6 volts. This effect increases with the magnitude of the diode's breakdown voltage. Should its breakdown voltage be below 5 volts, the voltage across the zener is now dependent upon the quantity of current flowing through it. Thus, if a zener diode is wired across a load resistor as shown in Fig. 6-12, and the voltage it is supposed to maintain across the load is 3.9 volts, it will maintain this voltage only if a fixed and specific amount of current is flowing through the reverse-biased diode. Voltage will change somewhat with changes in current. As a result, regulation of the voltage across R_L will be relatively poor with load and line variations.

Zeners with reverse breakdown voltages have 6 volts work in what is referred to as the *avalanche* region. If this voltage is below

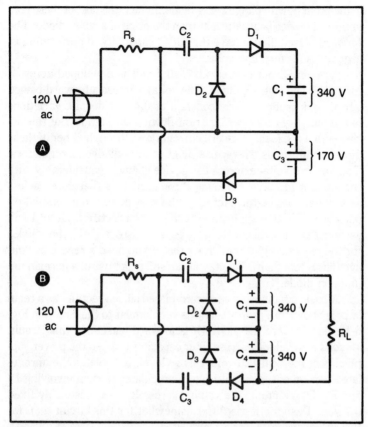

Fig. 7-9. Voltage multiplier. A. Voltage tripler. B. Voltage quadrupler.

5 volts as in our example, the diode is said to be operating in the *zener* region. Breakdown voltage remains relatively fixed despite current variations when the diode is operating in the avalanche region, but it varies to some degree when it is operating in the zener region. There is a cross between the two characteristics when the reverse breakdown voltage of the zener diode is between 5 volts and 6 volts.

A relatively fixed, low voltage is developed across the usual forward-biased diode when it conducts current. This voltage varies somewhat with the quantity of current flowing through the device, just as the reverse breakdown voltage varies somewhat with the current flowing through a zener diode operating in the zener region. Hence, there is little if any loss in performance if a forward-biased diode is used as the voltage-regulating device, rather than a low-breakdown-voltage zener diode. In fact, the cost of these conventional diodes is usually less than the price of a zener diode. The forward-biased diodes have the added capability of performing as regulators at very low voltages.

As pointed out earlier, about 0.7 volt is developed across a forward-biased silicon diode. The voltage across a forward-biased germanium diode is fairly constant at 0.4 volt despite variations in the quantity of forward current flowing through it. But this assumes that a sizeable forward current flows through either of these diodes at all times. Never use point contact diodes as regulators. The forward voltage across the device changes considerably with the amount of current flowing through it. It is therefore useless as a voltage regulator. Actually, the best of the conventional diodes to use in this application is the silicon rectifier. Here, I will assume that only this device is used, and disregard all other diodes for the moment. Just remember that if you need a regulator voltage that is less than 0.7 volt, you can still make use of a germanium junction diode.

A basic circuit using a forward-biased silicon rectifier as a regulator is shown in Fig. 7-10. This is very similar to the circuit in Fig. 6-12, except here the diode D_2 is forward biased. In this circuit, 12 volts is developed across filter capacitor C when the power supplied is at its ideal 120-volts rms ac level. R_S isolates the unregulated portion of the circuit from the regulated portion involving D_2 and R_L. D_2 maintains the voltage across R_L at a reasonably fixed 0.7 volt. Design criteria differ somewhat for this circuit than for the circuit using the zener regulator.

For example, let us assume that R_L = 70 ohms. Therefore, R_S must pass 0.7 volt/70 ohms = 10 mA of current for R_L. If D_2 were a zener diode, you would then say that it should conduct a minimum of one-tenth of the current that R_L conducts. But this is not a good criterion here. A forward-biased diode should conduct a minimum of 5 mA or better yet 10 mA so that the voltage across it and R_L should remain fixed to a reasonable degree when the diode current increases. Using 10 mA in this example, the total current through R_S is the current through the diode plus the current through R_L or 20 mA.

When the line voltage is at its 10% low, the 12 volts developed across C drops to about 10.8 volts. Because 0.7 volt is across D_2, the minimum voltage across R_S is 10.8 volts − 0.7 volt = 10.1 volts. Because the minimum desirable current is 20 mA or 0.02 A, R_S should be less than 10.1 volt/0.02 amp = 505 ohms. You

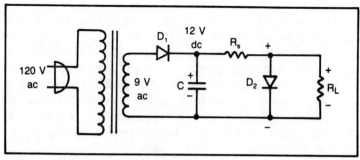

Fig. 7-10. Voltage regulator supply. The voltage across R_L is maintained at 0.7 volt.

should use a standard 470-ohm resistor.

Should the line voltage be at its peak, the voltage across C is 10% above the 12-volt average or 13.2 volts. Because this voltage less 0.7 volt is across R_S, the current flowing through that resistor is now 12.5 volt/470 ohms = 0.0266 A = 26.6 mA. Also, because 10 mA is used by R, 16.6 mA must flow through D_2. The voltage across the diode will increase somewhat above its ideal 0.7 volt, but it will be sufficiently stable to satisfy the requirements of many applications.

Power is dissipated by R_S and by diode D_2. The minimum power the resistor must be capable of dissipating is (12.5 volts) (0.0266 amp) or 0.332 watts. A resistor with at least double this dissipation rating should be used to minimize any chance of component breakdown. The diode must be capable of dissipating (0.7 volt) (0.01766 amp) = 0.0166 watt. It is very difficult to find a diode that will not meet this requirement.

In our previous discussion of zener-diode regulator circuits, you also determined the maximum current that must flow through that component. The zener diode must be capable of dissipating power equal to the product of two factors—the current flowing through it and its breakdown voltage. The power dissipation capability of any diode remains the same regardless of whether the current is flowing through it in the forward or in the reverse direction. It is always equal to the current flowing through it multiplied by the voltage across the diode.

As in the example with the zener-diode regulator, R_L can be considered as the component that varies rather than the supply voltage. Assume that R_L can vary from 7 ohms to 700 ohms. When it is 7 ohms, the current flowing through that resistor is 0.7 volt/7 ohms = 0.1 ampere. Since the diode should conduct a minimum of 0.01 ampere, the total maximum current flowing through R_S is the sum of the current through D_2 and R_L or 0.110 ampere. The voltage across R_S is 12 volts minus 0.7 volt or 11.3 volts. Consequently, R_S must be equal to or less than 11.3 volts/0.110 ampere = 102.7 ohms. Use a standard 100-ohm resistor. Using Equation 3-5, the power dissipated by R_S is (0.110 ampere)^2 (100 ohms) = 1.21 watts. A 2-watt resistor should consequently be used so that it will not get excessively hot when the current flowing through R_L is at its maximum.

The required minimum current for D_2 was determined under the condition that R_L be equal to 7 ohms. When R_L is a 700 ohms, the current that flows through it is 0.7 volt/700 ohms = 0.001 ampere, because 0.7 volt is across the component. Because 0.110 ampere is a fixed current that flows through R_S, and it must flow through both the diode and R_L at all times. Thus, 0.110 A – 0.001A = 0.109 A must flow through the diode when R_L = 700 ohms. This is more than ten times the current that flowed through the diode when R_L was 7 ohms. Consequently, the voltage across the diode will rise somewhat above 0.7 volt—maybe even up to 0.9 volt. In some circuits, this range of voltage variation is satisfactory despite the fact that regulation is poor. The power dissipated by the diode under these conditions, using Equation 3-4, is (0.9 volt) (0.109 amp) = 0.0981 watt. This is still quite low for the readily available devices.

Diodes in Series

Unlike the case with diodes used as rectifiers, zener diodes can be wired in series. A circuit of this type is shown in Fig. 7-11A. The regulated voltage is equal to the sum of the breakdown voltages of the diodes. In this example, it is 27 volts. Should the line vary by 10% and cause the dc voltage across C to vary by about this same 10%, the voltage across R_L remains fixed at the sum of the breakdown voltages of all zeners used in the circuit. Any combination of zener diodes can be used so long as the sum of their breakdown voltages add to the regulated voltage you require.

Forward-biased diodes can be wired into the same series arrangement to supply you with a specific regulated voltage. This combination is used primarily when a low, regulated voltage is required. In Fig. 7-11B, this voltage is three times the 0.7 volt that is across each diode or 2.1 volts. Here, as well as in the circuit using the zener diodes, you must be certain that each diode can tolerate the power it must dissipate. Just multiply the voltage across the

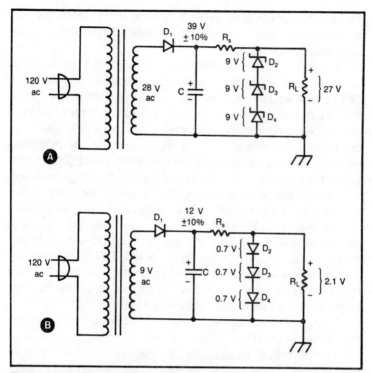

Fig. 7-11. Voltage regulator with diodes connected in series. A. Reverse-biased zener diodes. B. Forward-biased silicon diodes.

individual diode by the current flowing through it to determine the power it is dissipating. Check that its ratings are above the number determined from the arithmetic.

In all regulated circuits, the output voltage will be negative with respect to ground or the lower terminal of R_L if all diodes are wired into the circuit in the direction opposite to that shown in Fig. 7-11. This reversal must be done in the regulated and unregulated sections of the supply. If leads from capacitor C have specific polarities, the connections to this capacitor must also be reversed.

Modifying Regulated Voltages

Zener diodes have specified tolerance ranges. Inexpensive zeners are available in 10% categories. If you should buy one of the 12 volt zeners with this rating, its breakdown voltage can be up to 1.2 volts above or below the 12-volt specifications. Its breakdown voltage may be anywhere between 10.8 volts and 13.2 volts.

And you cannot complain to the distributor about the discrepency in breakdown voltage because you bought a diode with a ±10% tolerance specification.

More expensive zener diodes are supplied with ±5%, ±2%, and even ±1% tolerances. These zeners can differ from the specified 12 volts by 0.6 volt; from 11.4 volts to 12.6 volts; by 0.24 volt from 11.76 volts to 12.24 volts; and by 0.12 volt from 11.88 volts to 12.12 volts, respectively. But even if you buy a 10% zener, you can include one or more forward-biased diodes in the regulating circuit to compensate for a difference between the breakdown voltage of the zener diode and the actual regulated voltage you need.

Assume that because of the particular zener diode you are using in the circuit in Fig. 6-12 you get an output of 8.3 volts rather than 9 volts. But you need 9 volts across R_L. You can add 0.7 volt to the actual zener breakdown voltage by simply wiring a forward-biased silicon diode in series with the zener. This is shown in Fig. 7-12. The 0.7 volt across the forward-biased silicon diode is added to the 8.3 volts across zener diode D_2 for a total of 9 volts. This total voltage is then placed across load resistor R_L.

Using additional forward-conducting diodes in series with the one in the circuit, the regulated voltage can be increased in steps of about 0.7 volt. If, for example, you use three forward-biased diodes instead of only one D_3, the total voltage across these diodes would be 3×0.7 volt = 2.1 volts. Adding this to the 8.3 volts across the zener, a regulated 10.4 volts is now available for application across R_L.

If the zener diode's breakdown voltage is not exactly 0.7 volt less than the desired output voltage (or is not exactly a multiple of 0.7 volt less than the desired output voltage), you will have to compromise. For example, if the breakdown voltage of the zener diode is 8.1 volts, one additional forward-biased diode arranged as in Fig. 7-12 will put the regulated voltage at only 8.1 volts + 0.7 volt = 8.8 volts, even if 9 volts is specified for the supply. You may have to live with the available 8.8 volts. There is no problem if your equipment will work when 0.2 volt less than the ideal 9 volt supply is used as its power source. Otherwise, you can simply add another forward-biased diode in series with the zener circuit for a total available voltage of 8.8 volts + 0.7 volt = 9.5 volts. If your equipment will accept this excessively high voltage without breaking down, you could then use this voltage as an acceptable compromise.

An alternate solution involves the use of junction germanium

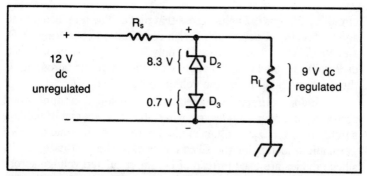

Fig. 7-12. Circuit to increase the regulated voltage above the zener-diode breakdown voltage. The total is applied across R_L.

diodes instead of silicon diodes. Now the addition to the zener diode voltage will be in 0.3 volt or 0.4 volt steps. These diodes may help you come closer to the desired voltage than silicon diodes. In fact, you may combine both types of diodes in the series circuit so that the total diode voltage plus the zener voltage supplies a more ideal source voltage.

The circuitry in Fig. 7-12 can be used if the zener voltage is less than the required regulated output voltage. But what if its breakdown voltage is greater than that voltage? Now you must reduce the voltage between the zener diode and the load to get the proper voltage for application across R_L. A way to accomplish this is shown in Fig. 7-13.

In Fig. 7-13, 9.7 volts is developed across zener diode D_2. Current flows in the forward direction through the diode in the D_3/R_L

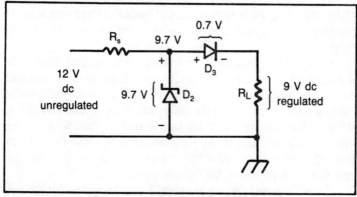

Fig. 7-13. Circuit to reduce the regulated voltage so that it is less than the zener-diode breakdown voltage. The total is applied across R_L.

circuit so there is 0.7 volts across that diode. The total zener voltage is developed across the series combination consisting of D_3 and R_L. With 0.7 volt across D_3, only 9.7 volts less the 0.7 volt, or 9 volts, is left for application across R_L. Voltage across R_L is the zener voltage minus the voltage across diode D_3.

As before, different voltages can be developed for application across R_L if several diodes are wired in series instead of using a single diode for D_3. Another method is to substitute one or more germanium diodes for the silicon device. In the first case, if two silicon diodes are used instead of D_3, the regulated voltage across R_L would be the 9.7 volts across D_2 minus 2(0.7 volt) = 8.3 volts. If three germanium diodes were used in its place, the regulated voltage would be about 9.7 volts minus 3 (0.4 volt) = 8.5 volts.

Should the desired output voltage be negative with respect to ground, all diodes must be wired in the reverse direction from that shown in the diagram.

Integrated-Circuit Regulators

Simple regulator circuits have been designed for many years using transistors in *feedback* arrangements. Feedback regulators utilize information about the voltage that is at the output of the regulator and across R_L. This information is fed back to a transistor circuit which passes the current to the output load. Voltage at the output of the regulator should be fixed at a specific magnitude. But it may shift slightly with line and load variations. Therefore, the circuit that passes current uses the information about the voltages the output to determine the amount of current it should pass. The quantity of current passed by the circuit flows through the load. Using this magnitude of current, the output voltage (equal to the product of this current with the load resistor) is reset to the desired level.

For example, assume the line voltage drops slightly so that the output voltage drops below what it would be under ideal regulated conditions. Information about this low voltage is fed back to the current pass circuit. Using this information, the circuit knows that it must pass more current to the load so that the voltage developed across it will increase to the required level. Because this all happens instantaneously, the load does not even know that the voltage has dropped below the desired magnitude.

Should the reverse condition exist and the voltage at the output of the regulator circuit increase above the desired level, the information fed back to the current-pass circuit forces it to reduce its current flow. The output voltage is thereby reduced to the required level.

Several ways to do this job have been incorporated into integrated circuits. These ICs have three terminals as shown in Fig. 7-14. The unregulated filter voltage is fed to the input of the IC, between the high input terminal and ground. A regulated voltage appears at the output of the IC, between the high output terminal and ground. The voltage and the polarity of the voltage at the output depends on the IC selected for the circuit. Current passed to the load is limited by the type of IC used. Do not let the IC overheat when drawing current. If you find that it is dissipating an excessive amount of power and getting very hot, mount the IC on a heat sink. This may be a vertically mounted metal plate that is about eight to ten times the size of the chip (IC). The metal plate will absorb heat from the IC and radiate this heat into the surrounding air so that the temperature will remain within reason.

The IC shown in Fig. 7-14, provides a positive output voltage with respect to ground. Other ICs are made which accept an unregulated negative input voltage and provide a regulated negative voltage at their output.

If the IC you have does not provide you with sufficient regulated output voltage with respect to ground, a diode can be used to increase that voltage. Just add the diode to the circuit as shown in Fig. 7-15. If the diode is a silicon device, it will increase the regulated output by 0.7 volt above ground or by this voltage above its original output. Two or more silicon diodes may be used to increase the output in multiples of 0.7 volt. Germanium junction diodes can be used instead of silicon devices to "up" the supply voltage in smaller increments. Should the output voltage be negative with re-

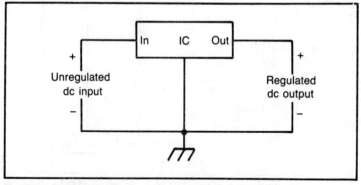

Fig. 7-14. Regulator using an integrated circuit.

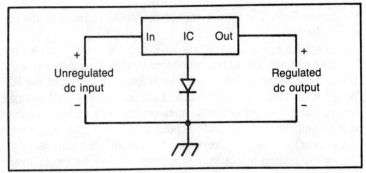

Fig. 7-15. The regulator output is increased by 0.7 volt because diode D is in the circuit.

spect to ground rather than positive, just reverse the direction in which the diodes are wired so that they will conduct in the forward mode.

Should you want to reduce the output voltage from the IC, a diode can be added between the output of the IC and the load as in Fig. 7-13. The regulated voltage from the IC is thereby reduced by the amount of the voltage developed across the forward-biased diode. In this case, as well as in the circuit where the voltage was increased, the amount of power the diode may dissipate safely must not be exceeded.

An example of a regulated supply with positive and negative output voltages is shown in Fig. 7-16. It is essentially the same circuit as in Fig. 7-6A but combined with two IC regulators. IC_1 provides a positive regulated output voltage with respect to ground,

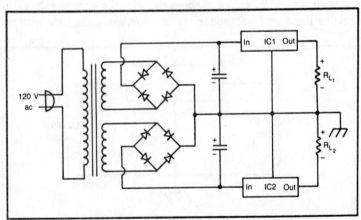

Fig. 7-16. Circuit providing positive and negative regulated output voltages.

while IC_2 provides a negative regulated voltage. If both halves of this circuit are identical except for the polarities across the loads, identical voltages with respect to ground appear across R_{L1} and R_{L2} with the polarities shown. This type of circuit is used widely to power medical equipment as well as to power some types of computers.

METERS TO MEASURE VOLTAGES

There are usually different voltages available from a power supply. Consider Fig. 7-2A. The primary produces 120 volts ac, while 9 volts ac is across each half of the secondary winding for a total of 18 volts ac. If there were a large filter capacitor across R_L, the voltage across the load and capacitor would be about 12 volts dc. A multimeter can be used to measure these voltages as well as the resistance of R_L. It can also be used to check if the diodes are in good condition.

In our earlier discussion of multimeters, I described meter circuitry used for measuring dc voltage and current as well as a circuit used in an ohmmeter. But you also need to be able to measure ac voltage.

Most meter movements respond only to dc current or voltage. To measure ac voltage using this type of movement, the ac must be converted to dc before being applied to the meter. A circuit to do this is shown in Fig. 7-17. Here, the ac passes through diode D_1 when the phase of the input voltage is such that the upper input terminal of the circuit is positive with respect to the lower terminal. By passing it through D_1, the ac is converted to dc. The converter acts as a half-wave rectifier circuit. The dc is applied to the meter movement, M. The amount that the meter pointer deflects is related to the amount of current flowing through R, D_1, and M which, in turn, is related to the magnitude of the ac voltage ap-

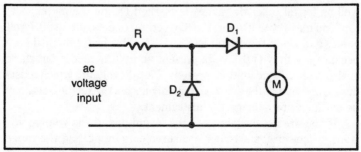

Fig. 7-17. Basic meter circuit used for measuring ac voltage.

plied to the circuit. The meter dial is calibrated to indicate the rms average of the applied ac voltage.

Diode D_2 is a path for leakage current. It keeps current from flowing through the meter during the portion of the ac cycle when the lower voltage input terminal is positive with respect to the upper terminal.

A commercially available multimeter is shown in Fig. 7-18. Two test leads are connected to the jacks at the lower corners of the panel. They are used for connecting the instrument to the circuit or component being tested. The black test lead is plugged into the jack marked " – " and the red lead is in the jack marked " + ." This polarity is important when using the instrument to measure dc voltage or current. To make these measurements, the red lead is connected to the more positive voltage in the equipment than is the black lead. When making ac voltage or resistance measurements, test leads may be connected in either direction to the voltage or resistor terminals.

In the first four counter-clockwise settings of the switch on the multimeter panel, the meter movement responds to dc voltages applied to the test leads. So if you are measuring a dc voltage, you must switch to one of these settings. The voltage noted on the panel is the maximum that can be applied to the test leads when the switch is in that position. Thus, if you want to measure 12 volts, you can not use the 10-volt setting because the meter pointer will go off scale. You can use any of the other settings, but maximum accuracy is obtained when you use the lowest setting that will accommodate the voltage being measured. In our case, we use the 50 DCV setting to measure 12 volts.

You can connect the test leads across R_L in Fig. 7-2A to measure 12 volts. The red lead from the instrument is connected to the upper terminal, while the black lead is connected to ground. The voltage is read by using the 0-50 scale. The scale used for this as well as for all other dc ranges is marked DC on the meter face.

You may try to use the 10 DCV setting to measure the 0.7 volt across a diode in the circuit in Fig. 7-2A. This isn't practical here because not only is 0.7 volts present across the diode, but there is also a reverse dc voltage present. This dc voltage, which exists during one-half of the ac cycle, not only upsets the reading but may be sufficient to damage the instrument.

If you are not certain as to the magnitude of the voltage you want to measure, start your measurements by putting the range switch into the highest setting. If the meter pointer indicates that

Fig. 7-18. Multimeter. (Courtesy EICO Electronic Instrument Company, Inc.)

the applied voltage is less than the highest voltage that the next lowest range will accommodate, then switch to that range. Keep switching to lower ranges until you find the lowest range that can accommodate the voltage you are measuring. Never switch to a lower range if the voltage read on the meter is greater than the voltage given on the panel. This is the maximum voltage that can be handled by that lower range.

Similar techniques can be used to measure ac voltages. But this time you must use the four maximum clockwise settings of the switch. The corresponding scale is the one marked AC on the face of the meter. If you connect the meter across the secondary winding of the transformer, you will measure about 18 volts. Use the 50 ACV setting to make this measurement. You will measure 9 volts across each half of the winding. The accuracy will be best if you use the 10-volt setting of the switch when making this measurement. Use the 250 ACV setting to check the line voltage or the voltage across the primary of the transformer. The reading should be about 120 volts.

Disconnect the power supply from the voltage source if you want to measure the resistance of R_L. Set the meter switch to one of the ohms ranges. Connect one lead from the instrument to the other lead so that there is 0 ohms between the leads. Adjust the 0-adjust control (not seen in the photo because this variable control is located at the left side of the case) for a 0-ohm reading at the upper end of the meter scale. Now connect the test leads across R_L. Note its resistance on the ohm's scale. If the meter pointer does not deflect, or if it deflects to one extreme end of the scale, try making the measurement on the alternate ohms range. For the final reading, use the range where the deflection of the pointer is closest to the middle of the scale. Multiply the number read on the scale by the multiplying factor noted in the meter setting being used for the measurement.

Note that in some instances, the resistor being measured may have other components wired across it. Some of these components may affect the reading. Readings are most accurate if one lead from the resistor is disconnected from the circuit before the measurements are made. Then measure the resistance by connecting the meter to the resistor's two terminals—one connected to the circuit and one hanging loose.

There is also a dc current range on the multimeter. If you want to measure the current flowing through R_L, first disconnect one lead of the resistor from the circuit. Connect the meter between the two disconnected leads to measure current flowing in R_L. Use the 0 to 10 DC scale and the group of calibration lines marked DC on the meter face make the readings. You must multiply this by 10 because the current range is from 0 to 100 mA and the scale printed on the meter face is from 0 to 10.

Although this multimeter has only one current range, other instruments have more. Because current ranges are seldom used, the one range on this instrument is usually all that is needed.

Ac current cannot be measured directly with this or any other analog multimeter—only digital meters have this capability. To measure ac current using this instrument, you must break into the ac circuit. Insert a small known resistor of a few ohms between the two points where you broke into the circuit. Then, using the multimeter, measure the ac voltage across the resistor. Using Equation 2-1, the current flowing in the circuit is equal to the ac voltage read on the multimeter divided by the resistance.

In Fig. 7-2, you would measure the ac current in the primary of the transformer by disconnecting one of the leads from the supply, from one terminal of the primary winding on the transformer (after first having unplugged the unit from the power line, of course). Connect a known low-ohmage resistor to once again complete the circuit between the line and terminal for the primary winding. Reconnect the circuit to the power line. Measure the ac voltage across the resistor inserted between the line and transformer. Using the measured voltage and the known resistance, you can determine the ac current by applying Equation 2-1.

A simple meter is shown here. Many more complex instruments are readily available from EICO and other manufacturers. Operation and use of these instruments, are very similar to the procedures presented here. It is best that you follow the procedures detailed in the manuals supplied with the instrument you have, to utilize that instrument's maximum capabilities.

Measuring Diodes

In the discussion of diodes, I noted that they conduct when the anode is positive with respect to the cathode and do not connect when a voltage of the opposite polarity is applied to their terminals. Using this information, an ordinary ohmmeter can be applied to determine if a diode is in good shape. The only requirement is that more than 0.7 volt must be present across the test leads of the instrument when it is set to measure resistance. This is usually the case with analog meters such as the one shown in Fig. 7-18. However, this is seldom the case with digital multimeters. Therefore, this test procedure applies only when most types of analog meters are involved.

Start by disconnecting one lead of the diode in question from the circuit. Set the instrument so that it is on one of the ohmmeter ranges. Connect the test leads to the two terminals of the diode and note the resistance. Next reverse the connections of the test leads to the diode. The reading should be either higher or lower. If the two readings are the same, the diode may be defective. If they differ, the diode is likely in good operating condition.

If your initial test shows that the diode is defective, do not discard it. Make the same measurements using other ohms ranges. If you get two different readings on any one of the ranges when the lead connections are interchanged, then the diode is probably good. This is because this difference in resistances will not appear when extremely high or extremely low ohmmeter ranges are used.

Protecting the Meter Movement

If too much voltage or current is applied to the instrument when the switch is set to a particular range, some of the components in the multimeter may be damaged. If the ruined component turns out to be a resistor or diode, it is a relatively inexpensive undertaking to get a resistor or diode for replacement purposes. But if the excess voltage or current damages the meter movement, replacement may be so expensive that it is sometimes cheaper just to "junk" the entire instrument and buy a new one.

Circuits have been developed to protect the meter movement. One of the most commonly used circuits is shown in Fig. 7-19.

The pointer on the meter movement deflects because current, I_M, flows through the coil of the meter assembly. This coil has a resistance, R_M. Voltage V_M is developed across the coil because of the current. Using Equation 2-3, this voltage is equal to $V_M = I_M R_M$. Because the meter movement can safely tolerate a specific maximum current, it can likewise tolerate a specific maximum voltage. This current and voltage is usually several times greater than the current and voltage required to deflect the pointer to full scale. Diodes D_1 and D_2 are wired across the meter movement to keep this voltage within safe levels.

If the maximum voltage the particular meter movement can tolerate is 0.7 volt, two silicon diodes can be wired across it as

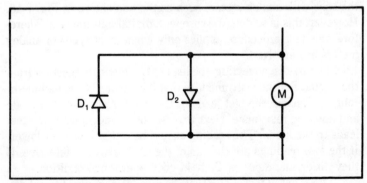

Fig. 7-19. Diodes are used to protect the meter movement.

shown for protection purposes. When excess positive voltage is at the upper terminal of the movement with respect to its lower terminal, diode D_2 conducts. Up to 0.7 volt can be across D_2. Because it is across the meter movement, D_2 limits the voltage across it to a maximum of 0.7 volt. It does not affect the normal performance of the meter because much less than this 0.7 volt is required to deflect the pointer to full scale. The diode will not conduct when these low voltages are present.

Should the excess voltage applied to the meter movement be opposite to the polarity just noted, diode D_1 conducts to keep the voltage across the movement to within the safe 0.7 - volt level. If the voltage is kept to within safe levels, so is the current, because the current is related to the limiting voltage by Equation 2-1.

Leakage

Many modern pieces of equipment have line cords with three wires. Two leads are for supplying the 120 volts ac. The third lead gets connected to a terminal in the wall outlet that is wired to ground. At the equipment, the two leads from the 120-volts ac are connected to provide the required power. But the ground lead is connected to the chassis. This is to keep the chassis at ground despite any magnetic fields or electric current leakage inside the equipment. If you touch the chassis or metal cabinet with one hand, you cannot get an electric shock because it is connected to ground. This is true even if you touch a radiator or water pipe with the other hand. You will not get an electric shock even if there is some current leakage from the line to the chassis.

A circuit using this safety arrangement is shown in Fig. 7-20. This type of circuit is always used in medical equipment. The grounding arrangement frequently involves line cords connected to air conditioners, refrigerators, washing machines, and any other equipment housed in a metal cabinet or in equipment constructed with exposed screws that are connected to a metal chassis.

Two measurements should be made to ascertain whether the ground lead from the cable is fulfilling its function properly.

1. Disconnect the power plug from the line. Check the resistance between the ground terminal (usually the center terminal) of the plug and the chassis or equipment cabinet. The resistance should be less than 0.2 ohms, but up to 0.5 ohms is acceptable in most instances.

2. Next measure the ac leakage current. To do this, discon-

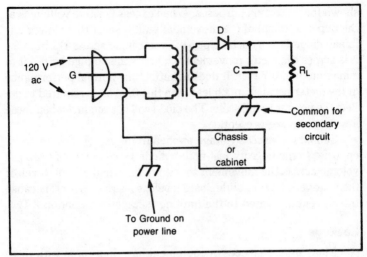

Fig. 7-20. Three-conductor cable supplies power to the circuit.

nect the power plug from the line. Next disconnect the ground line in the power cable from the equipment chassis. Connect a known resistor between the disconnected line and the point on the chassis from which the line was disconnected. Reinsert the power plug into the ac wall outlet. Measure the ac voltage across the resistor. Determine the current that is flowing in the line and resistor through Ohm's law, Equation 2-1. This leakage current should be quite low—in the microamperes region. Information as to the allowable limits for any piece of equipment should be obtained from the manufacturer. The limit is frequently set at less than 200 μA.

If you should find that the resistance or current measured does not conform with the desired limits, check the ground wire connection in the line cord. It must be making good contact with the ground terminal in the plug that is wired to the cable if the resistance is to be low. Also, make sure that all screws on the cabinet and chassis are tight. Be sure that the plug is clean and that there is no spillage of fluids inside the equipment. The line cord itself may cause a problem if it is old and corroded. If you cannot find and correct the fault, the manufacturer of the equipment should be consulted.

Chapter 8

Bipolar Junction Transistors

Circuits using only ordinary junction diodes as active devices cannot provide voltage or current gain. Transistors, however, can perform as amplifiers in conjunction with the other components in the circuit. Various arrangements using transistors are used to take the signal voltage, current, or power present at the input of a circuit and magnify it. A magnified version of the input signal will then be present at the output of the amplifier.

There are other amplifying devices besides transistors. A number of years ago, before the advent of transistors, the vacuum tube was the most important amplifying device. Now it is essentially an outdated component, although it is still used in some special applications. Two types of transistors have superseded the vacuum tube. These are the bipolar junction transistor and the field-effect transistor. The bipolar junction transistor will be discussed in this chapter, while details about the field-effect transistor will occupy our attention in Chapter 9. Another type of transistor, those in the point contact group, may still be found in equipment designed in the 1950s. This device is no longer in current use.

THE JUNCTION TRANSISTOR

Junction diodes are constructed from two different types of semiconductors—an n-type and a p-type. A depletion region exists between these slabs. The structure of the junction transistor elaborates on that of the diode. It consists of three semiconductor slabs and two depletion regions.

One type of transistor consists of two sections of n-type material with a very thin piece of p-type material placed between the two n-type slabs. A second type of transistor has the opposite structure with a thin n-type slab sandwiched between two p-type slabs. Using the structure with the slabs to identify the transistor groups, the first type with two n-type slabs and one p-type slab is referred to as an *npn transistor*. The second type is a *pnp device*. Structural diagrams of both types of transistors are shown in Fig. 8-1.

The schematic representation of the two groups of junction transistors is shown in Fig. 8-2. The semiconductor that is sandwiched between the outer two slabs is referred to as the *base*. The outer two slabs are the *emitter* and the *collector*. (See Fig. 8-2.) The emitter lead is denoted by an arrow. It points away from the base in the schematic of an npn transistor and points towards the base in the drawing of a pnp device. As for the diodes, the arrow here indicates the direction of conventional current flow and *not* the direction of the electron current.

Using a diagram of the structure of the pnp transistor, you can determine how a transistor works. In Fig. 8-3A, the three semiconductor slabs are shown. There are two depletion regions in this device. One is at the base-to-emitter junction and second is at the base-to-collector junction. Two depletion areas exist because there are two p-n junctions. There is only one of these junctions in a diode so it has only one depletion region.

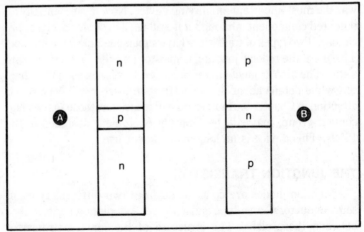

Fig. 8-1. Structure of junction transistor. A. Npn type. B. Pnp type.

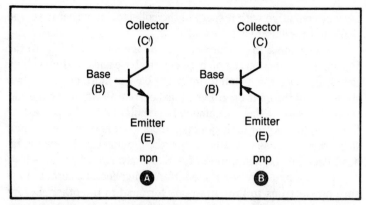

Fig. 8-2. Schematic representation of transistors. A. Npn. B. Pnp.

Now note the diagram in Fig. 8-3B. The dc voltages applied to the base and collector are both negative with respect to ground. This polarity is used only when a pnp device is involved. Voltages of opposite polarities are applied to the transistor when an npn device is used in the circuit. In either case, E_{CC} is higher with respect to ground that the voltage at the base of the transistor.

With no voltage applied to the transistor slabs, the depletion regions shown in the figure are formed. Now apply $-E_{CC}$ volts to the collector. With the collector, C, at a negative voltage with respect to the emitter, E, no current can flow between these terminals. This is because of the depletion regions between these semiconductors. The same condition would exist even if the emit-

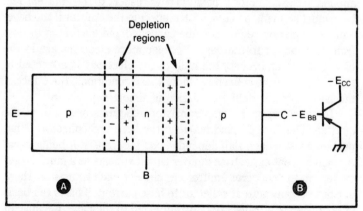

Fig. 8-3. The pnp transistor. A. Structure of transistor including the depletion region. B. Schematic showing relative voltages with respect to ground, applied to terminals for conventional use.

ter were negative with respect to the collector. In either case, the depletion regions resist current flow.

Now connect the negative base voltage supply, $-E_{BB}$, to the B lead. The depletion region between the base and emitter is eliminated. Because the n-slab forming the base is more negative than the p-slab of the emitter, conventional current flows from the emitter to the base. It was the same voltage polarity that helped establish a flow of current in the junction diode. But here current cannot flow from the base to the collector: the n-p junction is reverse biased because the magnitude of $-E_{BB}$ is always less than $-E_{CC}$.

The voltage between the emitter and collector exerts a pressure on electrons to flow from one terminal to the other. Before the base voltage is applied, two depletion regions exist and current could definitely not flow between the emitter and collector. By applying voltage to the base, you eliminate one barrier to current flow—the depletion region between E and B. But there is still a second barrier—the depletion region between B and C.

Once current flows from the emitter to the base, an excess number of holes builds up in the narrow base region. They have nowhere to go and so force their way into the collector slab.

The holes present in E are transferred to the base. These holes are attracted to the negative collector. Because of this attraction, the base-collector depletion region is almost eliminated. The bulk of these holes travels to the collector. A lesser number passes into the base lead.

If an npn transistor is used rather than a pnp device, voltage applied to the collector is positive with respect to the emitter. The base must be made positive with respect to the emitter if the base-emitter depletion region is to be cancelled and current is to flow from the emitter to the base. Because more electrons are in the base than it can absorb, the bulk of the electrons is attracted to the highly positive collector. Under these conditions, the effect of the base-collector depletion region is almost nonexistent.

In considering the npn transistor, electrons originate at the emitter. They are divided between the base and collector. What portion goes to each slab depends upon the transistor being used. A small current in the base-emitter circuit appears as a much larger current in the collector-emitter circuit. For each transistor, there exists a certain ratio of collector to base current. This is the basis of operation of a transistor. Current is fed to the base. Because of the base current, current flows in the collector. The collector current base is equal to the current magnified by a certain factor. This "magnifying factor" is the ratio of the total current from the emitter that flows in the collector to the amount of current that is in the base. When the base current is magnified by this amount, the produce gives the current flowing in the collector. The ratio of the collector current to the base current is the current gain of the transistor. The bipolar junction transistor is basically a current amplifier.

Here, I used an npn device in the description of the current amplifier. The same facts apply to the pnp device, except that here the holes from the emitter divide up between the base and collector rather than the electrons. By making the base negative with respect to the emitter, a small current flows in the base. Because of this current, a larger current flows in the collector-emitter circuit if the collector is made negative with respect to the emitter. The collector current is related to the base current by this current-magnifying factor.

In the discussion that follows I will concentrate primarily on the npn transistor. Just keep in mind that the facts also apply to the pnp device. In the latter device, the applied voltages must be reversed in polarity from those applied to the npn transistor. Later, however, I will show how a pnp transistor is used in conjuction with an npn device in one circuit.

The following circuit description and operation will be based on the use of silicon transistors. When the base-emitter junction is forward biased, there is 0.7 volt across this junction. This is the same as the voltage across the forward-biased silicon diode junction. Therefore, if a germanium transistor is to be used in a circuit, the only difference between that device and the silicon transistor that must be considered is the base-emitter voltage. For the germanium transistor, it is assumed that this voltage is around 0.3 volt or 0.4 volt.

CURRENT GAIN OR LOSS

Junction transistors are basically current amplifying devices. You produce voltage gain by sending the currents through resistors or resistances of appropriate magnitudes. The ratio of voltages developed across the resistances due to the flow of the currents in these components is the voltage gain of the transistor circuit. This is illustrated in more detail below.

The current-gain characteristic of a particular transistor can be described by either one of two numbers. One number relates the amount of current flowing in the collector of the transistor, to the amount of current flowing in its base. This ratio is referred to as the transistor's *beta*, or β. Stated mathematically,

$$\beta = \frac{I_C}{I_B}$$

Eq. 8-1

where I_C is the collector current and I_B is the base current.

The beta of different transistors ranges from less than 10 to more than 800. If the beta of a particular transistor is 10 and 10 mA of current is fed to its base, its collector current is $\beta \, I_B = 10(10 \text{ mA}) = 100 \text{ mA}$. If you need 100 mA of collector current from this transistor, you must feed an $I_B = I_C/\beta = 100 \text{ mA}/10 = 10 \text{ mA}$ to the base. All this can be determined from Equation 8-1.

In a similar fashion, if the transistor you are using has a beta of 800 and you feed 100 μA to the base, the collector current is $800(100 \, \mu\text{A}) = 80,000 \, \mu\text{A} = 80 \text{ mA}$. Just before, if you want a current of 80 mA in the collector circuit of the transistor, you must apply 80 mA/800 = 0.1 mA = 100 μA to the base of the device.

Current gain in either instance is obviously equal to the ratio of the collector to the base current. Equation 8-1 is essential for determining the performance of a transistor and should be remembered.

The beta of every transistor is different. Because transistors are so variable, no two transistors will have identical betas—if they did, it would be a fluke. To establish some sort of order, similar types of transistors are grouped in accordance with their betas. (By "similar type," I mean devices that have the same characteristics in all respects except for current gain.) For example, consider the five groups of transistors that fall into the same category and have been assigned numbers 2N5824 through 2N5828. The group with number 2N5824, has betas that range from 60 to 120; the 2N5825 group has a beta range of 100 to 200; the 2N5826 group has a beta range of 150 to 300; the 2N5827 group has a beta range of 250 to 500, and the group numbered 2N5828 has a beta range of 400 to 800. All these transistors are identical in every way except for the beta category into which individual devices may fall. They have the same maximum voltage rating, the identical power dissipation capabilities, identical limits on the highest frequency that they can amplify, and so on.

If the beta of a particular device is within a specific range, the manufacturer of the transistor must assign the number to it that is used for transistors in that beta range. Thus, if the beta of a transistor in the 2N5824 to 2N5828 group is 115, it may be assigned either number 2N5824 or 2N5825 and not any other number. If it had a beta of 225, it would have assigned the number 2N5826.

A second number which describes the current capabilities of the transistor is its *alpha*. The Greek letter α is used to represent this current ratio. Alpha relates the collector current to the emitter current by the formula

$$\alpha = \frac{I_C}{I_E} \qquad \text{Eq. 8-2}$$

All transistor current flows in the emitter. It is split between the base and collector with the bulk of the current from the emitter flowing into the collector. Therefore, less current flows in the collector than in the emitter. The alpha, the ratio of the collector to emitter currents, is consequently less than 1. It is almost always very close to 1, usually between 0.98 and 0.999. If a specific current is fed to the emitter, less current than I_E flows through the collector.

Because alpha and beta are both related to the currents in the transistor, and all transistor currents are related to each other, alpha and beta are related to each other. If you only know the alpha of a particular transistor, you can determine its beta by using Equation 8-3.

$$\beta = \frac{\alpha}{1 - \alpha} \qquad \text{Eq. 8-3}$$

Similarly, if you need to know the alpha of a transistor but only know its beta, you can determine this gain through Equation 8-4.

$$\alpha = \frac{\beta}{1 + \beta} \qquad \text{Eq. 8-4}$$

CIRCUITS WITH TRANSISTORS

There are basically three circuit arrangements in which transistors are used. All other circuits are simply variations on these. Each circuit type is designated by the transistor terminal used as the reference point for the dc or for the ac signal voltages at the other two terminals. In our initial discussion, this reference terminal was chosen as the emitter. Here, I will start our analysis by using the base as this reference point.

Common Base Circuit

The basic circuit used in the *common-base* arrangement is shown in Fig. 8-4. The base terminal is the common or reference point. A negative voltage is applied to the emitter with respect to the base so that this junction is forward biased. Voltage between the collector and base is of such polarity as to reverse bias this junction. Resistor R_E in conjunction with E_{EE} sets the emitter current, I_E. The arrow at I_E shows the direction in which conventional current flows in the base-emitter circuit.

Because current flows through the base-emitter circuit 0.7 volt is across this junction. The total voltage in this section of the circuit outside the transistor is $E_{EE} - 0.7$. The emitter current is equal to this voltage divided by the total resistance in this circuit. Here, this resistance is only R_E.

Emitter current is equal to the sum of the base and collector currents. Because very little base current flows compared to the collector current, the collector current, I_C, is essentially equal to the emitter current, I_E. With the current at the input equal to the current at the output, the current gain, I_C/I_E, is just about equal to 1. It is actually equal to the alpha of the transistor, which is close to 1. Voltage at the collector with respect to the base is equal to E_{CC} minus the voltage developed across R_C due to I_C, or $E_{CC} - I_C R_C$.

Transistor circuits are usually designed to amplify an ac input voltage rather than dc current. In this type of arrangement, the ac is superimposed on the dc. The dc is used to establish specific quiescent emitter and collector currents. These *bias currents* should set the idling voltage (voltage before ac is applied to the circuit) at the collector at one-half of E_{CC}. Because the collector voltage, V_C, is equal to $E_{CC} - I_C R_C$, V_C should be idling at about $E_{CC}/2$.

Now lets place an ac voltage, V_{In}, in series with the existing dc voltage in Fig. 8-4. This is illustrated in Fig. 8-5. Here, it is as-

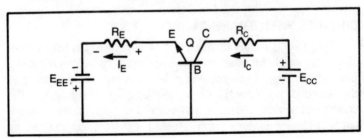

Fig. 8-4. Basic common-base circuit.

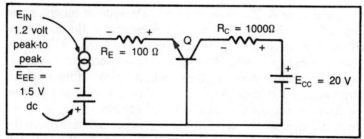

Fig. 8-5. Common-base ac amplifier.

sumed that the emitter supply voltage is fixed at 1.5 volt and that an ac with a 1.2 volt peak-to-peak level is superimposed on the dc. This combination is applied to the base-emitter circuit.

Adding ac sinusoidal voltage to dc voltage can be accomplished graphically. An example of this is shown in Fig. 8-6. For this illustration, it is assumed that the ac input signal varies from 0 volt to a peak of +1 volt and to a crest of −1 volt. The dc voltage, E_{EE}, is chosen at different fixed levels.

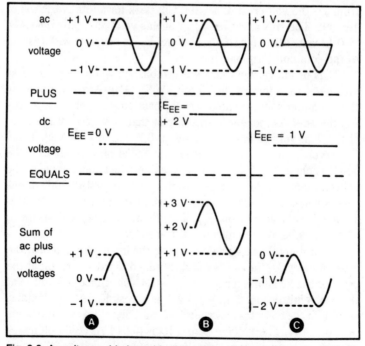

Fig. 8-6. Ac voltage added graphically to a dc voltage. A. Dc is 0 volt. B. Dc is +2 volts. C. Dc is −1 volt.

219

In Fig. 8-6A, E_{EE} is set at 0 volt. The total ac superimposed upon this is thereby equal to and due entirely to the ac voltage. There is no dc to add to it because the dc is at 0 volt. In Fig. 8-6B, the dc level is +2 volts. With +1 volt to −1 volt peak-to-peak of ac superimposed upon this, the sum of the peaks of the ac signal with dc is +2 volts dc plus +1 volt ac or +3 volts, and the sum at the crest of the ac signal is +2 volts dc plus −1 volt ac or +1 volt. In Fig. 8-6C, the identical situation exists except that now the dc voltage is negative and equals −1 volt. Adding the ac to this gives a peak of −1 volt dc, plus 1 volt ac, or 0 volt and a crest of −1 volt dc, plus 1 volt ac, or −2 volts. In all instances, the total output voltage varies between the crest and peak during different parts of the cycle. It follows the cyclic variations of the pure ac at the input.

Getting back to the circuit in Fig. 8-5, let us use the numbers there to illustrate the addition of the ac signal to the dc bias voltage and discuss how the circuit performs its amplifying function. The ac-input-signal voltage is given here as 1.2 volt peak-to-peak. The peak voltage is one-half of this or 0.6 volt, so the ac varies from a −0.6 volt crest to a +0.6 volt peak for a total variation of 1.2 volt. The rms average, E_{RMS}, of this voltage is 0.6 volt/1.414 = 0.42 volt. Also note that, as before, E_{EE} is negative with respect to the common terminal, the base.

Bias or idling conditions are established when there is no ac voltage at the input of the circuit, or E_{In} = 0. At this time, only the dc emitter supply voltage less the base-emitter voltage is across R_E, the only component in this circuit that is outside of the transistor. This voltage is −1.5 volt minus 0.7 volt, or −0.8 volt. Thus, −0.8 volt is across R_E. Emitter current is now equal to −0.8 volt/100 ohms = −0.008 A. Because this is just about equal to the collector current, the voltage across R_C is (−0.008 A) (1000 ohms) = −8 volts. The polarity is as shown at the resistor. When this voltage is subtracted from E_{CC}, 20 volts minus 8 volts, or as much as 12 volts, is at the collector of transistor Q.

Now apply the ac signal voltage to the input. When it is at the crest of its cycle, the −1.5-volt dc is added to the −0.6 volt at the crest of the ac signal for a total of −1.4 volt. Since 0.7 volt is at the base-emitter junction, −1.4 volt is across R_E. At this instant, current in the collector and emitter circuits are −1.4 volt/100 ohms = −0.014 A. The voltage across R_C is (0.014 A) (1000 ohms) = 14 volts, so the voltage at the collector of Q is (20 volts) − (14 volts) = 6 volts.

When the applied ac voltage is at its positive peak in the cycle, the $+0.6$ volt ac is added to the $-1.5=$ volt dc, for a total of 0.9 volt. Subtracting the 0.7 volt that is at the base-emitter junction, -0.2 volt is across R_E. I_E and I_C are at this instant equal to (-0.2 volt)/(100 ohms) $= -0.002$ A. The voltage across R_C is (0.002 A) (1000 ohms) = 2 volts. Now the voltage at the collector is 20 volts minus the 2 volts across R_C or 18 volts.

Voltage calculated to be at the collector must never drop below 0 volt or exceed the supply voltage—20 volts in this example. If either of these happen when ac voltage is at the input, the voltage across R_C or between the collector and base of the transistor would differ in shape from the voltage applied to the input. A good amplifier provides an output signal that looks exactly like the input. They differ only in that the output and input may have different overall magnitudes due to amplification by the transistor circuit.

Dc voltage at the collector is E_{CC} minus the voltage across R_C. If the collector current (current through R_C) were so high as to make the voltage across R_C greater than E_{CC}, our calculations would show that a negative voltage is at the collector of the transistor. But this cannot happen because collector voltage cannot drop below the zero volt at the base reference terminal under any circumstances.

Now let's see what happens when an ac signal is applied to the input. Assume that during a portion of the ac cycle, the sum of the ac voltage and dc bias voltage is sufficient to increase I_C to the level where calculations will show that the voltage across R_C exceeds E_{CC}. During this interval, the voltage at the collector of the transistor would theoretically drop below zero. But this cannot happen. It remains at its 0-volt minimum. The output will not follow the cyclic variation of the input in this portion of the cycle. It remains fixed at 0 volt. This portion of the waveform is distorted.

The same is true if during a portion of the cycle the sum of the ac input voltage and the dc bias voltage adds to a magnitude that is below zero volt. During this portion of the cycle, I_E and I_C would stay at zero, putting the voltage across R_C at zero volt. Now the collector voltage remains fixed at E_{CC}. During this interval, the output voltage is a distorted version of the input signal.

To avoid these problems or minimize the chances of this happening, the dc idling current is adjusted so that the collector bias voltage is at about one-half of E_{CC}. Now the output voltage can vary over its maximum range from zero to E_{CC} volts. R_E is also chosen so that the emitter and collector currents are limited to values where the $I_C R_L$ voltage never exceeds E_{CC} even when the ac input voltage is at its peak. The total current must be less than E_{CC}/R_L or less than $(E_{EE} - 0.7)/E_E$.

Getting back to our example, note that the voltage across R_C varies from 14 volts to 2 volts, while the ac input signal varies from -0.6 volt to $+0.6$ volt. Therefore, the peak-to-peak output was 14 volts $-$ 2 volts = 12 volts while the peak-to-peak input was 0.6 volt $-$ $(-0.6$ volt$)$ = 1.2 volt. The voltage gain of the circuit is the output voltage divided by the input voltage or 12 volts/1.2 volt = 10. But also note that the ratio of R_C to R_E is 1000 ohms/100 ohms = 10. You can therefore conclude that the ac voltage gain of the common-base circuit is equal to R_C/R_E.

Using Equation 3-5, you know that the power across the output circuit is $I_C^2 R_C$, and the power at the input circuit is $I_E^2 R_E$. Because I_E is just about equal to I_C, the power gain or ratio of the power at the output to the power at the input is $I_C^2 R_C/I_E^2 R_E$ = R_C/R_E. Because this ratio is the same as the voltage gain, the power gain and voltage gain are identical.

When the ac voltage at the emitter is at its peak with respect to the base, the voltage at the collector is also at its peak with respect to the base. The crests of the ac signal also exist at the same time at the emitter and collector. You can therefore conclude that the ac output from a common-base circuit is in phase with the ac input to that circuit.

Common-Emitter Circuit

The common-base arrangement is used primarily as a high-frequency amplifier. Common-emitter circuits are used more frequently than are the common-base circuits. These arrangements are applied to most applications in the audio band. Common-emitter circuits are also used in some very high-frequency equipment.

The positive voltages applied to the base and collector in the common-emitter arrangement are with respect to the emitter. This is shown in Fig. 8-7 where the emitter terminal of the transistor is connected to ground. The base supply, E_{BB}, in conjunction with R_B, sets the idling current, I_B, in the base. It is equal to E_{BB} minus the 0.7 volt between the base and emitter divided by R_B, the only resistance in this portion of the circuit. This current, multiplied by beta, is equal to the idling collector current, I_C. Because I_C flows through R_C, voltage is developed across that resistor in the collector circuit. As was true in the common-base circuit, it is also desirable here to let the voltage at the collector terminal in this circuit

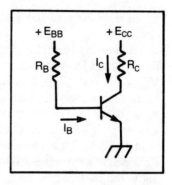

Fig. 8-7. Common-emitter circuit.

idle at $E_{CC}/2$ volts. To do this, the idling voltage across R_C must be equal to half of the supply voltage, or $I_C R_C = E_{CC}/2$. Because I_C is related to I_B, I_B is set by the designer of the circuit to equal the desired I_C divided by beta. I_B can, of course, be set by using the proper base resistor, R_B, in the circuit.

Ac signal voltage is usually fed through a capacitor to the base and applied between the base and emitter to the transistor. The input voltage varies the base and hence collector current in step with the cyclic variations of the ac. A magnified version of the ac input voltage is developed across the collector resistor by the ac collector current flowing through it. Output voltage is fed from the collector through a capacitor to a load of some type. This is shown in Fig. 8-8. The load is drawn here as a resistor. A resistor has also been added between the emitter and ground to simulate the usual circuit. Instead of having a separate supply for the base current,

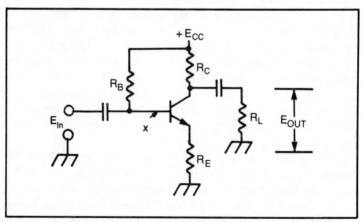

Fig. 8-8. Common-emitter ac amplifier. The beta of the transistor is an important factor in determining the operation of the circuit.

223

E_{CC} is used here as the supply for both the collector and base circuits.

Base idling current is established by the circuit involving E_{CC}, R_B, V_{BE}, and R_E. The voltage in this current is $E_{CC} - V_{BE}$, where $V_{BE} = 0.7$ volt. Current in R_B is I_B, while current in R_E is βI_B or I_E. In determining the base current, therefore, you can consider R_E as being in series with the base, replacing the lead at which arrow "x" is pointing. But before putting it there, R_E must be multiplied by beta. This is because the actual voltage developed across R_E is due to a current equal to beta multiplied by the current flowing in the base circuit. So in order for the same voltage to be developed across the resistor when moved to the base circuit as was developed across it in its original location, R_E must be multiplied by beta before being relocated. As a result, the total resistance in the base circuit is $R_B + \beta R_E$. Base idling current is (E_{CC} − 0.7)volts/($R_B + \beta R_E$)ohms = I_B. Idling collector and emitter currents, I_C and I_E, are both about equal to beta times I_B.

As indicated, I_C should be of such magnitude as to establish a collector-emitter voltage and idling voltage at the collector equal to about one-half of E_{CC}. This is achieved when $I_C(R_C + R_E) = E_{CC}/2$.

The ac input signal adds a variable current to the dc idling current in the base. It is amplified by the transistor so that a magnified version of the ac input is across R_C. When the voltage at the base is at its peak in the cycle, the collector current is also at its peak. Because the voltage across R_C is the product of the collector current with R_C, this voltage is also at its maximum level. Subtracting the peak voltage across R_C from E_{CC}, you will find the voltage at the collector of the transistor at this instant. This is its minimum with respect to ground. Because the output voltage is at a minimum when the input voltage is at a maximum, you can conclude that the input and output voltages are 180° out of phase. This is the situation that exists with common-emitter circuits. Despite the fact that the input and output voltages are out of phase, the magnitude of the ac voltage at the output of the transistor is usually greater than it is at the input.

For example, let $R_C = 5000$ ohms, $R_L = \partial$ ohms, $R_E = 500$ ohms, $E_{CC} = 22$ volts, and the beta of the transistor be equal to 100. The total resistance in the collector-emitter circuit is $R_C + R_E = 5000$ ohms + 500 ohms = 5500 ohms. If the idling voltage across the transistor is to be one-half E_{CC} or 11 volts, I_C must be equal to 11 volts/5500 ohms = 0.002 ampere. Now 11 volts will be across the total resistance in the collector-emitter circuit or 5500 ohms, as well as between the collector and emitter of the transistor.

If 0.002 A (or 2 mA) is to flow in the collector circuit, that current divided by beta must be in the base circuit. Because I_C = 0.002 A, I_B = 0.002 A/β = 0.002A/100 = 0.00002A = 2×10^{-5} A. Therefore, the resistance in the base circuit must be (E_{CC} − 0.7)/I_B = 21.3 volts/2 $\times$ 10^{-5}A = 10.65 $\times$ 10^{-5} ohms = 1,065,000 ohms. R_E appears in the base circuit as if it were beta times what it actually is or 100 $\times$ 500 ohms = 50,000 ohms. Because this resistance plus R_B is the total resistance in the base circuit, R_B must be equal to 1,065,000 ohms − 50,000 ohms or 1,010,000 ohms. A minor 1% error will occur if you let R_B equal 1,000,000 ohms or 1 megohm.

Current gain of this circuit is equal to the beta of the transistor. If 0.02 mA of current is fed to the base, beta times 0.02 mA is at the collector. In this example, with beta equal to 100, 100 $\times$ 0.02 mA = 2 mA flows through the collector (and emitter) circuit.

In order to determine the ac voltage gain, you must take into account the *ac emitter resistance* that is present inside the transistor. It is equal to 0.026/I_E, with I_E, the idling current, expressed in amperes. Thus, if I_E = I_C = 2 mA = 0.002A, the ac emitter resistance, r_e, in this example is 0.026/0.002 = 13 ohms. When ac flows in the circuit, this resistance is usually significant. As far as the ac is concerned, the total resistance in the emitter circuit is R_E + r_e. This is equal to 500 ohms + 13 ohms = 513 ohms. The dc resistance of the emitter circuit remains fixed at R_E. Here it is 500 ohms.

Using 513 ohms as the total ac resistance in the emitter circuit, lets determine the ac voltage gain of the circuit. Assuming E_{In} has a peak of 0.5 volt and a crest of − 0.5 volt, the peak-to-peak variation of the input signal is 1 volt. This variation is applied to the series circuit consisting of the base-emitter junction, r_e and R_E. Because the input voltage is in the base circuit, it sees the 513 ohms in the emitter multiplied by the beta of the transistor or 51,300 ohms. The base current swing due to the E_{In} variation of 1 volt is 1 volt/51,300 ohms = 1.95 $\times$ 10^{-5}A. The transistor current is not cut off even when the swing of E_{In} is down to − 0.5 volt at its crest because the base is biased by the dc to idle at 2 $\times$ 10^{-5} mA. The negative crest of the input signal tends to reduce the current at that instant of the cycle, but it will not counter the dc idling current enough to stop the transistor from conducting.

The ac collector current swing is beta multiplied by the base current swing. Because beta is 100, the collector current swing is $1.9 \times 10^{-5}A \times 100 = 1.9 \times 10^{-3}A$. The peak-to-crest voltage developed across R_C due to the collector current swing is $(1.95 \times 10^{-3}A)$ (5000 ohms) = 9.75 volts. The ratio of the output voltage across R_C to the input voltage is 9.75 volts/1 volt = 9.75. This is the ac voltage gain, A_v, of the transistor circuit.

By looking at the circuit, you can find a very simple way of determining the voltage gain. It is approximately equal to $R_C/(R_E + r_e)$. Here, it is 5000 ohms/513 ohms = 9.75. This is the same gain number calculated using the longer procedure above.

The power at the input to the transistor is the rms base current multiplied by the applied voltage, or $I_b E_{rms}$. At the output, the ac power is the rms collector current multiplied by the rms voltage developed across R_C, or $I_C E_{rms}$. The power gain of the circuit is the ratio of the output power to the input power or $I_C E_{OUT}/I_b E_{rms}$. Because I_C/I_b is equal to beta and E_{OUT}/E_{rms} is equal to the voltage gain, the power gain, G, is beta times the voltage gain or βA_v. In this example, it is 100(9.75) = 975.

When dealing with practical circuits, three additional factors usually require some consideration. One is the signal source. Like the battery, the ac signal source is not perfect. It has some internal resistance. If this resistance is substantial, it must be take into account when calculating the gain of the overall circuit.

The second factor is the dc supply. In all circuits, whether the supply is a battery or a converted ac arrangement, a capacitor is connected across it. Because a capacitor passes ac and behaves as an open circuit for the dc, the supply is a short circuit for the ac. The negative terminal of the $+E_{CC}$ supply in Fig. 8-8, is at ac ground (unless shown otherwise). As far as the ac or signal voltage is concerned, the top terminal of R_C and R_B are both at ground.

A load resistor, R_L, is wired across R_C through a capacitor. The capacitor behaves as a short circuit for the ac, letting it pass freely. The end of R_C that is connected to $+E_{CC}$ is at ac ground. One end of R_L is also connected to ground. R_L is effectively in parallel with R_C through the coupling capacitor. The output load for the ac signal is the equivalent resistance of the parallel combination of R_L and R_C. R_L is usually so large as to be negligible when compared to R_C, but not always. The loading by R_L on the output circuit is the third factor of concern to us here.

So now let's modify Fig. 8-8 and change it to the circuit shown in Fig. 8-9. C1 conducts the ac from E_{In} to the base, and C2 con-

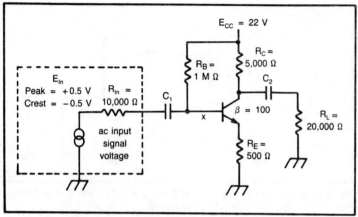

Fig. 8-9. Practical common-emitter circuit.

ducts the ac from the collector to R_L. These capacitors are again essentially short circuits for ac and open circuits for dc. Because of this, they do not let dc from the transistor circuit enter the E_{In} supply or the R_L load. They also do not let the resistance of the supply or load upset the dc bias and idling conditions established by E_{CC} and resistors R_B, R_E, and R_C.

In Fig. 8-9, all components in the dc circuit are identical to those used in the previous example. All idling currents are unchanged. $I_B = 2 \times 10^{-5}$A, and $I_C = 2 \times 10^{-3}$A. Because r_e is the ac emitter resistance of the transistor it does not affect dc currents or bias voltages.

The ac input "sees" resistor R_B between the base and ground. The resistance between the junction of R_B and the base is $\beta(R_E + r_e) = 51,300$ ohms. The connection at the base is effectively at ground, for all resistors and resistances between the emitter and the ground, $R_E + r_e$, have been moved to the left side of the base. This leaves the emitter at ground. Therefore, E_{In} sees the 51,300 ohms as if it were a resistance between the R_B/C1 junction and ground. In the actual circuit, this resistance is shown as wired between the base terminal and ground.

The 51,300 ohms appears to shunt R_B, the 1 MΩ base resistor. Thus, the resistance at the input of the transistor is equal to the parallel combination of both items, or (1,000,000 ohms) (51,300 ohms)/(1,000,000 ohms + 51,300 ohms) = 48,797 ohms, or approximately 49,000 ohms. Considering this resistance in conjunction with the 10,000 ohm resistance of E_{In}, a voltage divider is formed and Equation 3-2 applies. This circuit is shown in Fig. 8-10A.

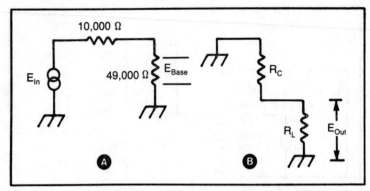

Fig. 8-10. Effects of the ac portion of the circuit in Fig. 8-9. A. Equivalent of input circuit. B. Equivalent of output circuit.

The total ac input voltage, E_{In}, is applied to the divider and base of the transistor through its internal 10,000 ohm resistance. The load presented to this circuit is 49,000 ohms. Using the voltage divider equation, E_{BASE} = 49,000 ohms/(49,000 ohms + 10,000 ohms) = 0.83 multiplied by E_{In}. If the peak and crest voltages supplied by E_{In} are 0.5 volt, the peak and crest voltages at the base are 0.83 × 0.5 volt = 0.415 volt for a peak-to-peak voltage swing of 2 × 0.415 volt or 0.83 volt.

The ac load resistance in the collector circuit consists of R_C in parallel with R_L. This is because the upper terminal of R_C is effectively at ac ground and C2 behaves as if it were a short circuit for the signal voltage. This is shown in Fig. 8-10B. The equivalent resistance of the combination is (5000 ohms) (20,000 ohms)/(5000 ohms + 20,000 ohms) = 4,000 ohms.

Gain is approximately equal to the ratio of the impedance in the collector circuit to the impedance in the emitter circuit. In this example, the gain of the transistor circuit is 4000 ohms/513 ohms = 7.8. So with peak-to-crest voltage of 0.83 at the base of the transistor, 7.8 × 0.83 volt = 6.5 volts is the peak-to-crest voltage across R_C and R_L. With this as the output voltage, the gain of the overall arrangement is 6.5 volts/1 volt = 6.5. This gain figure takes into account the resistors in the dc portion of the circuit as well as the input impedance of the supply and the ac load due to the presence of R_L. Note how much lower this gain is than the gain in the original example when R_{In} and R_L were not being considered. There, the gain was taken as being simply $R_C/(R_E + r_e)$ = 5000/513 = 9.75. This figure is 9.75/6.5 or 1.5 times the actual gain of the circuit.

Common-Drain Circuit

There are times when voltage gain is not important and only the input and output impedance of a circuit are of concern. In Fig. 8-9, losses would be minimized if E_{In} saw a large load impedance and if R_C were much smaller than R_L. In the first case, the magnitude of R_{In} would be insignificant because practically all of the input voltage would be developed between the base and emitter. Only a very small voltage would be lost because of R_{In} if this situation existed. As for the output, all the voltage would be across R_C, while the shunting or loading effect of R_L would be negligible. But these conditions cannot be achieved solely through the use of the common-emitter circuit. A *common-drain* circuit would have to be added to help fulfill these goals. This arrangement is also known as an *emitter follower*. The circuit with ac inputs and outputs is shown in Fig. 8-11.

The collector is connected directly to the $+E_{CC}$ supply. As a result, the collector is at ac ground through the $+E_{CC}$ supply. Input from the ac source is therefore applied to the base-collector junction of the transistor. Output voltage is developed across R_E. Because the collector is at ac ground, the output voltage is between the emitter and collector. Ac input and output voltages are therefore considered here as being with respect to the collector.

The dc portion of this arrangement has requirements similar to those of the common-emitter circuit. Here, however, one-half of the E_{CC} supply voltage should be across R_E because there is no resistor in the collector circuit. In the diagram, I set R_E equal to 5000 ohms and E_{CC} equal to 20 volts. Therefore, $E_{CC}/2 = 10$ volts should be across R_E to satisfy the same idling current requirement

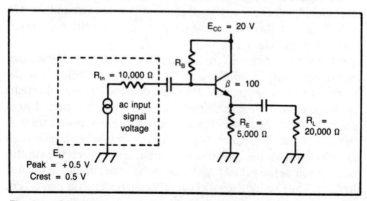

Fig. 8-11. Common-drain circuit with ac input and output.

229

noted as in the common-emitter circuit. To accomplish this, the dc idling collector (or emitter) current must be 10 volts/5000 ohms = 0.002A = 2mA. If the beta of the transistor is equal to 100, the base current must be I_C/β = 2mA/100 = 2×10^{-5}A. This is the bias current that flows through R_B.

Idling voltage across R_B is equal to E_{CC} minus the 0.7 volt across the base-emitter junction, and that difference minus the voltage across R_E. This voltage is therefore equal to 20 volts − 0.7 volt − 10 volts = 9.3 volts. If the current through R_B is 2×10^{-5}A, that resistor must be equal to 9.3 volts/2.3 × 10^{-5}A or about 400,000 ohms.

In this circuit, R_E is larger than it was in the common-emitter arrangement. Because R_E = 5000 ohms and R_L = 20,000 ohms, the ac resistance in the emitter circuit is (5000 ohms) (20,000 ohms)/ (5000 ohms + 20,000 ohms) = 4000 ohms. This resistance, when reflected into the base portion of the circuit, is 100 × 4000 ohms = 400,000 ohms. It is much larger than the reflected resistance in the common-emitter circuit, because there only 500 ohms between the emitter and ground. Here, r_e is not important in determining the reflected resistance because it is usually negligible when compared to R_E. The ac and dc emitter resistance of the emitter-follower differ by only a small amount.

The resistance in the base circuit is due to the parallel combination of R_B and the reflected resistance from the emitter circuit. This is (400,000 ohms) (400,000 ohms)/ (400,000 ohms + 400,000 ohms) = 200,000 ohms. Using this as the resistance presented by the base circuit to the ac signal, the peak and crest voltages at the base can be determined from the equivalent circuit in Fig. 8-12A. From Equation 3-1, you will find these voltages are 0.5 volt(200,000 ohms)/(200,000 ohms + 10,000 ohms) = +0.476 volt and −0.476 volt. This is so close to E_{In} that the losses in this portion of the circuit are negligible.

In the output circuit, the resistance seen between the emitter and ground is R_E in parallel with the resistance reflected from the base circuit and divided by beta. This division by beta rule stems from the earlier demonstration that the resistance reflected from the emitter into the base circuit appears there as the resistance multiplied by beta. Therefore, the opposite is true when a resistance is reflected from the base circuit. It appears between the emitter and ground as the actual resistance in the base circuit but divided by beta. The resulting parallel combination is shown in Fig. 8-12B.

As far as the base circuit is concerned, the ac resistance is R_B

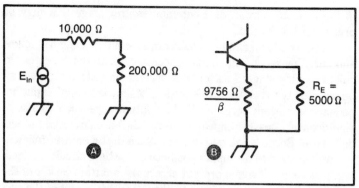

Fig. 8-12. Effects of the ac portion of the circuit in Fig. 8-11. A. Equivalent of input circuit. B. Equivalent of output circuit.

in parallel with R_{In}. It is obvious that R_B is between the base and ground because its upper terminal is at $+E_{CC}$. A 0-ohm ac resistance exists between $+E_{CC}$ and ground. As for R_{In}, it is the only resistance of the ac supply. Because it has one end at ground, the resistance is effectively connected between the base and ground. Thus, the total base-to-ground resistance of this parallel circuit is (R_B) $(R_{In})/(R_B + R_{In})$ = (400,000 ohms) (10,000 ohms)/(400,000 ohms + 10,000 ohms) = 9756 ohms. This resistance, divided by beta, is between the emitter and ground as shown in Fig. 8-12B. Because this circuit uses a transistor with a beta of 100, the resistance reflected into the emitter circuit is 9756 ohms/100 or 97.56 ohms. Since R_E = 5000 ohms, the actual resistance in the emitter circuit is (97.56) (5000)/(97.56 + 5000) or about 96 ohms.

Output voltage is developed across the 96-ohms located between the emitter and ground. This resistance feeds the output voltage developed across it to R_L. Because R_L is much larger than the 96 ohms, it has a negligible effect on the ac signal voltage developed in the dc portion of the emitter circuit and will not affect the gain of the circuit.

Voltage across R_E is slightly less than the voltage applied between the base and emitter. Therefore, the voltage gain of this circuit is just a trifle less than 1.

Current in the emitter circuit is just about equal to that in the base circuit multiplied by the beta of the transistor. Consequently, the power gain of this circuit, equal to the product of the current and voltage gains, is also just about equal to beta.

This circuit is very useful. Its high input impedance presents a negligible load to a signal source as well as to a resistor that is in the collector circuit of a common-emitter stage that may be preceding this circuit.

In conventional circuits, one common-emitter stage will follow another. The second stage may excessively load the collector circuit of the first stage. To avoid this, a common-collector circuit can be inserted between the two stages. Not only is its high-input impedance important here, but its low output impedance is also of significance. This low impedance feeds the base circuit of the second stage. Because the output impedance of the emitter follower is low compared to the input resistance of the base circuit it is feeding, just about all signal present across the emitter resistor of the emitter-follower will be across the base-emitter circuit of the succeeding stage.

COUPLED STAGES

As I just mentioned, there can be more than one transistor in a single circuit. An arrangement showing a two stage amplifier using two common-emitter transistor circuits is in Fig. 8-13. Signal is fed to Q_1 and is amplified. Because the load in its collector circuit is R_1 in parallel with that presented by the base circuit of Q2, the voltage gain of Q_1 is the equivalent resistance of the parallel combination of these resistances divided by R_{E1}. Similarly, the load in the collector circuit of Q_2 is R_L in parallel with R_2. Gain of Q_2 is the resistance of this parallel combination divided by R_{E2}. The gain of the overall circuit is the product of the gains of the two transistors. This gain is reduced somewhat by the presence of R_{In}.

Note that a capacitor, C_2, is used between the two transistors. It lets the ac signal pass from the output of Q_1 to the input of Q_2,

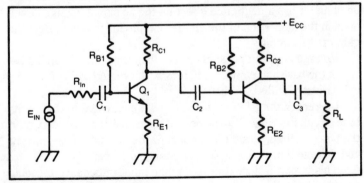

Fig. 8-13. Two-stage amplifier.

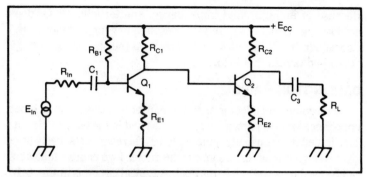

Fig. 8-14. Direct-coupled amplifier.

Note that a capacitor, C_2, is used between the two transistors. It lets the ac signal pass from the output of Q_1 to the input of Q_2, but it isolates the dc voltages of the two devices. In this way, both transistors can be biased independently. Capacitors, unfortunately, do not pass the low frequencies as well as they do the highs. Direct-coupling methods are therefore used where very low frequencies or even dc must be amplified. An arrangement showing this is shown in Fig. 8-14. Here, current flows through R_1 to the collector of Q_1 as well as to the base of Q_2.

Emitter followers were described as presenting a high impendance to the input signal and a low impendance to the load. They are shown as being used in this fashion in conjunction with a voltage amplifier in Fig. 8-15. Here, the ac input is fed to Q_1. Because it is an emitter follower, its high resistance in the base circuit is a negligible load on the ac input signal. The output from Q1 remains fixed at almost the magnitude of the ac input voltage. This voltage is fed to the base circuit of Q_2 and amplified by that stage.

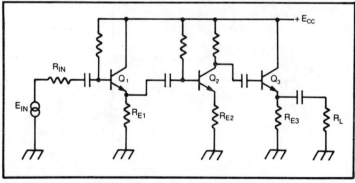

Fig. 8-15. Some methods of putting emitter followers to good use.

Because of its high input impedance, emitter-follower Q_3 does not load the output of Q_2. It does present a low impedance to R_L. Consequently, R_L has little or no effect on the size of the voltage developed across this load.

DARLINGTON PAIRS

The common-collector arrangement produces a specific input impedance when a particular resistor is used in the emitter circuit. This impedance is usually quite high, but there are instances where an even higher input impedance is desirable. Two emitter followers can be used to accomplish this goal. A typical circuit is in Fig. 8-16 and it is referred to as a *Darlington pair* or *Darlington amplifier*.

The resistance seen between the base and ground at Q_2 is equal to the resistance between the emitter and ground of Q_2 multiplied by its beta. In a similar fashion, the base-to-ground resistance at Q_1 is its emitter-go-ground resistance multiplied by its beta. Since the emitter-to-ground resistance of Q_1 is equal to the beta of Q_2 times its emitter-to-ground resistance, the resistance presented at the base of Q_1 is the emitter-to-ground resistor at Q_2 multiplied by the product of the betas of the two transistors. This resistance is, of course, in parallel with R_B.

In a similar fashion, the resistance presented by the input circuit to R_L is equal to the equivalent resistance of R_{In} in parallel with R_B divided by the product of the betas of the two transistors.

NOISE

All amplifier circuits have problems with interference. In the

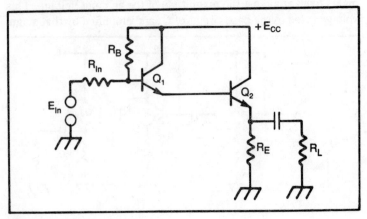

Fig. 8-16. Darlington circuit.

ideal case, only the input signal would be amplified and only this signal would appear at the output. But this aim is seldom achieved. There is a lot of random audio noise or radio frequency interference present in the air surrounding the amplifier. An amplifier will pick up some of the noise and amplify it. If the noise level is substantial, it will be heard in the background behind the amplified signal.

Transistors are not perfectly noise-free. Like others, they generate some background noise. The noise developed at the output of an amplifier is related to the quality of a particular transistor used in a circuit as well as to the quality of the passive components. This is in addition to the noise picked up in the air around the amplifier.

In order to determine the amount of signal at the output of a circuit with respect to the noise, a signal-to-noise ratio is used. First the output signal voltage is measured. Then the signal is shorted or removed from the input of the amplifier so that no signal is amplified. Only the noise voltage remains at the output and is measured. This relationship between the signal and noise voltages at the output is known as the *signal-to- noise ratio*. The voltage ratio is usually expressed in terms of *decibels* or dB where the larger number is the most desirable figure. Table 8-1 shows how decibels are related to voltage ratios.

For example, assume the two voltages measured at the output of an amplifier are 10 volts of signal and 1 volt of noise. The signal

Table 8-1. Decibel Notation for Voltage Ratios.

dB	Voltage Ratio
0	1:1
1	1.1:1
2	1.3:1
3	1.4:1
6	2:1
7	2.2:1
10	3.2:1
12	4:1
13	4.5:1
14	5:1
17	7:1
20	10:1
40	100:1
60	1000:1

to noise voltage ratio is 10:1. Checking this in Table 8-1, the ratio can be expressed as 20 dB.

Specifications of amplifiers usually list the signal-to-noise ratio in dB. Let us say the specification calls for 40 dB. By checking the numbers on the chart, you can see that the signal voltage is 100 times greater than the voltage of the noise.

But what are you to do if the ratio in dB is not listed on the chart? You can simply apply the rule that when you multiply numbers of voltage ratio; you must add the numbers in dB.

Assume that the specification calls for a signal-to-noise ratio of 65 dB. Looking at the table, you see only 60 dB. There is no 5 dB to add to it, for a total of 65 dB. But you can add 2 dB and 3 dB to the 60 dB for a total of 65 dB. The voltage ratio is the product of the ratios indicated by the dB numbers. Consequently, the signal to noise voltage ratio is (1.3/1) (1.4/1) (1000/1) = 1820/1 = 1820:1.

In a similar fashion, if you know the voltage ratio, you can divide it by other numbers to determine the ratio in dB. If this ratio happens to be 11,000:1, you can separate the number into the product of three numbers. These are $11,000 = 1000 \times 10 \times 1.1$. The sum of the equivalent numbers in dB is 60 dB + 20 dB + 1 dB = 81 dB. If the original ratio cannot be divided into an exact number of terms, you can estimate the actual number of dB represented by that ratio, through use of approximations to the actual multipliers.

Db notation is not only useful only as an indicator of the signal-to-noise ratios. It is also useful in applications where gain of an amplifier must be different at different frequencies. For example, the frequency response of a phonograph amplifier which amplifies the output from a magnetic cartridge, is not flat. Its output should be about 14 dB less at 10,000 Hz than it is at 1000 Hz. To see if the amplifier is performing properly, you would feed 1000 Hz to its input and note the output voltage. Then you would feed 10,000 Hz to its input instead of the 1,000 Hz, and once again note the output voltage. The ratio of the two output voltages when expressed in decibels, should be 14 dB. Using Table 8-1, you can determine that the voltage ratio should be about 5:1. Filters are used to establish this gain-difference situation.

FILTERS

There are basically four types of filters used in different applications. One type, the *low pass*, is designed to reduce the output voltage or amplifier gain of the high-frequency end of a band with respect to the low frequencies. An alternate type is referred to as a *high-pass* filter. It reduces gain at the low frequencies compared to the higher frequencies. There are also *bandpass* filters that permit a narrow group of frequencies to pass relatively easily through a circuit while all other frequencies are attenuated, and *band-rejection* filters that attenuate a narrow group of frequencies while letting all others pass freely. In this discussion, I will be discussing primarily high- and low-pass filters.

Modern high- and low-pass filters are constructed from resistors and capacitors. Circuits using these components are in Fig. 8-17. The basic filters are shown in Fig. 8-17A and B. If the *rolloff* or loss in gain at one end of the band with respect to the second end is not fast enough, then several of these filters can be used in the circuit. Arrangements of this type are shown in Fig. 8-17C and D.

In some applications, the rolloff will start at some low frequency and continue to a specific high frequency. The gain should remain fixed at that high frequency. A circuit doing this is shown in Fig. 8-17E. The way the gain varies in each of these arrangements is shown in Fig. 8-18. The letter associated with each graph is for the circuit with the same letter in Fig. 8-17.

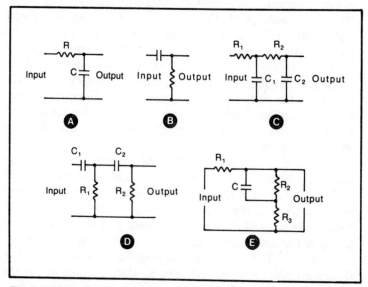

Fig. 8-17. Filter circuits. A. Low pass. B. High pass. C. Fast-rolloff low pass. D. Fast-rolloff high pass. E. Filter with rolloff over a limited frequency range.

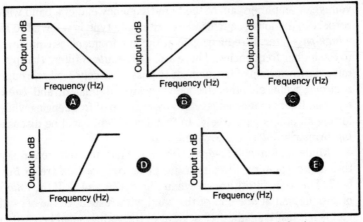

Fig. 8-18. Frequency response of filters in Fig. 8-17. The labels for the response curves are identical with the ones used for the filters.

Many different filter types and arrangements exist. To establish the frequency characteristic curve needed for a phonograph playback amplifier, the RC circuit in Fig. 8-17A can be placed between two amplifier stages. This is shown in Fig. 8-19A.

The required frequency-response curve for a phonograph record reproduction can be more accurately approximated through use of a filter network in a feedback loop. This is shown in Fig. 8-19B. Here, the signal is fed to the amplifier. All frequencies are amplified equally of Q_1 and Q_2 so that the gain over the entire band is the same. Ideally, you would be able to see this *linear* response (unchanging gain over the entire band) at the collector of Q_2 as a fixed output at all frequencies. However, the output does vary with frequency. If the R_1-R_2-C_1 network were not in the circuit, the response would be linear and gain would be the same at all frequencies. But it is there to reduce the gain of the overall circuit at the high frequencies.

Output from the collector of Q_2 is fed through a resistor/capacitor network back to the emitter of Q_1. Because of the capacitor in the feedback circuit, this network passes the high frequencies more readily from the output to Q_1 than the low frequencies. All frequencies passed by this network appear across R_{E1}. Because of C_1, there are more high frequencies here than low frequencies.

In Fig. 8-13, the input signal was applied between the base of Q_1 and ground. Only the portion of E_{In} that appeared across the input of the transistor between its base and emitter was amplified. Since a portion of the signal was across R_{E1}, R_{E1} limited the amount of signal that was between the base and emitter. Therefore, R_{E1} was an important factor in determining the gain of the transistor Q_1, which is equal to R_1/R_{E1}. The ac voltage developed across R_{E1} due to E_{In} affects the amount of signal appearing across R^1.

Now apply this to the circuit in Fig. 8-19B. A portion of the output is fed back and appears across R_{E1}. This voltage must be subtracted from E_{In} in order to determine the voltage that is between the base and emitter of Q_1 for amplification. The signal voltage left for application to the base-emitter circuit of Q_1 and amplified is E_{In} minus the voltage across R_{E1}. That difference is

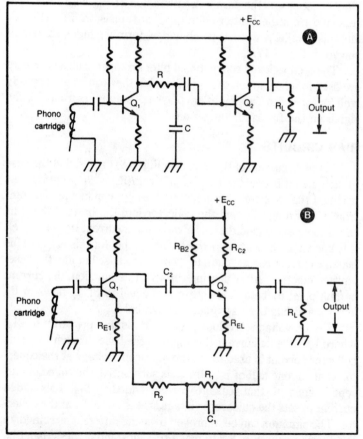

Fig. 8-19. Amplifier for signals from phono cartridge. A. Approximated response by using the circuit in Fig. 8-17A. B. More accurate response using a feedback circuit.

across the base-emitter input terminals of the transistor. It is this difference voltage that is amplified by the circuit.

Because more high frequencies are fed back from the output than are low frequencies, fewer high frequencies remain between the base and emitter at the input terminals of Q_1 after the feedback voltage is subtracted from E_{In}. All frequencies are amplified linearly by the circuit. Thus, the same voltage ratio is at the output of the amplifier as was established at the input due to the presence of the $R_1/R_2/C_1$ filter network in the feedback circuit.

It is also desirable that the very low frequencies be attenuated for proper reproduction of the recorded signal. This is accomplished by the combination of C_2 with the resistance at the base of Q_2. This resistance is the parallel combination of R_{B2} with resistance reflected from the emitter circuit of that transistor. Together, C_2 and this parallel resistance combination, form the high-pass circuit shown in Fig. 8-17B.

The feedback and conventional filter arrangements shown here are used for phonograph preamplifiers, but they can be used equally well for tape recorder amplifiers playback amplifiers, rolloff of the signal on the fm band, and so on.

BIAS CIRCUITS

In the discussion of the circuit in Fig. 8-8, I noted that the base idling current is equal to $(E_{CC} - 0.7)$ volt/$(R_B + \beta R_E)$ ohms. This is true if the base-collector junction does not conduct any current from E_{CC} through R_C and the collector into the base. If this current should exist, (and there is always some current flowing here), it is known as the *leakage current*. The symbol for this is I_{CB0}. This leakage current has nowhere to go from the base but into the base-emitter junction. After passing through this junction, the current is multiplied by beta. This current is equal to βI_{CB0}, flows in R_E and R_C and has been assigned the symbol I_{CE0}.

The dc voltage developed across R_C is due primarily to βI_B, where I_B is the dc current flowing into the base through R_B. This collector current is used to set the quiescent voltage at the collector at about one-half of E_{CC}. If I_{CE0} is substantial, the collector current is equal to that leakage current added to βI_B. This added current upsets the quiescent voltage that is normally at the collector. The situation can be improved by using different circuits than the one shown in Fig. 8-8 to make I_{CE0} substantially less than βI_B. A number of different bias circuits have been developed. Some of these are shown in Fig. 8-20. These can be used to reduce I_{CE0} for

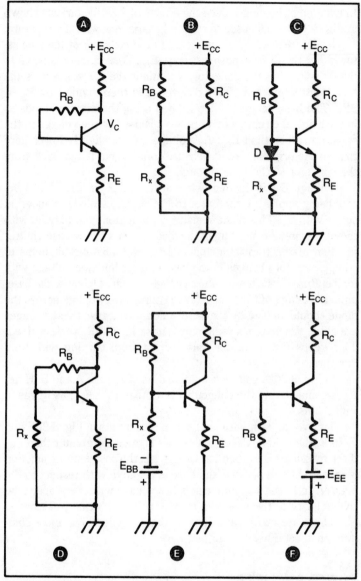

Fig. 8-20. Bias arrangements used in electronic equipment.

fixed values of I_{CB0}. The diagrams shown here are based on the common-emitter circuit, but the circuits can be used to provide equivalent characteristics in any of the other situations.

Transistor idling and leakage currents drift with temperature.

In addition to helping reduce the effects of I_{CBO}, the circuits shown in Fig. 8-20 are also used to reduce I_C variations with temperature.

In Fig. 8-20A, R_B is connected to the collector of the transistor rather than to the power supply, E_{CC}. Base current is equal to the dc voltage at the collector, V_C, minus the 0.7 volt across the base-emitter junction. This difference is then divided by R_B + βR_E. The base current is determined using the same methods as in the original circuit, except now V_C is used as the voltage in the equation rather than E_{CC}. Ac voltage gain is slightly lower here than in the original circuit because some signal is fed back from the collector to the base through R_B.

Voltage E_{CC} is divided between R_B and R_x in Fig. 8-20B before being applied to the base. In Fig. 8-20C, a diode is added in series with R_x. This diode-resistor combination is in parallel with the series circuit consisting of the diode formed by the base-emitter junction of the transistor and resistor R_E. This circuit helps to compensate for changes in the base-emitter junction voltage with temperature. If the temperature change is such as to cause the base-emitter voltage of the transistor to change, the voltage across the diode should change by the same amount. Because these changes are alike, the voltages remaining across R_x and R_E remain fixed. This keeps the transistor bias current from varying with temperature.

In Fig. 8-20D, you see a combination of circuits A and B. Here, R_B is connected to the collector rather than to $+E_{CC}$, as it was in Fig. 8-20B.

Two voltage supplies are used in the circuits in Fig. 8-20E and F. In E, the voltage developed across R_x must be greater than E_{BB} if the transistor is to conduct. Otherwise, the base-emitter junction is reverse biased because the base is negative with respect to the emitter. In F, the V_{EE} source supplies voltage with the proper relative polarity to the base and emitter to turn on that junction.

There are many variations on these circuits. The most common ones have been described here.

POWER AMPLIFIERS

Power gain is one of the characteristics that I discussed earlier for the various transistor circuits and is an important factor. Power is frequently required at the output of an amplifier. Power is necessary to drive a loudspeaker, feed signal to a head on a tape recorder, light a bulb when an amplifier is turned on, and so on.

Single-Ended Amplifier

In the circuit in Fig. 8-8, E_{In} is applied to the input of the amplifier. E_{In} may be audio, rf, or anything else, but in our discussion it will usually be audio. All the other applications can be derived from this analysis of audio circuits. Here, I will consider E_{In} as the voltage at one frequency of the voice or music signal applied to the amplifier. By convention, it is usually taken to be 1000 Hz. E_{In} sees a resistance equal to R_B in parallel with βR_E. Applying Equation 3-6, you know that the power supplied by the source to the amplifier, is E_{In}^2 divided by the resistance it sees at the input to the transistor. Here, E_{In} represents the rms voltage fed to the input, E_{rms}, rather than the peaks or crests of the ac voltage in cycles. These peaks and crests were used in our discussion above. The rms voltage will be used below.

Output current at the frequency of E_{rms} flows through R_C and R_L. If R_L is a loudspeaker, you need at least one-fourth of a watt to drive it. (This assumes that you are driving a small and efficient loudspeaker.) But R_L is usually a low resistance, frequently about 4 ohms. It is much lower in resistance than the R_C and R_E used in the dc portion of the transistor circuit. Being much lower than R_C, you can assume that the resistance in the collector circuit (R_L in parallel with R_C) is equal to R_L. Because the ac voltage gain of the stage is the ratio of the resistance in the collector circuit to R_E or R_L/R_E, you know that this voltage gain is very low, probably much less than 1. Power gain is beta times the voltage gain, so it too is low. This is true even if beta is equal to 500. A transistor circuit of the type drawn in Fig. 8-8 cannot, under normal circumstances, supply enough power to drive a loudspeaker.

One solution is to increase the resistance or impedance in the collector circuit. To do this, an audio transformer is used rather than R_C. The load is wired across the output or secondary of the transformer. This arrangement is shown in Fig. 8-21.

When discussing transformers, I noted that the impedance presented to one winding is related to the impedance in the second winding by the square of turns ratio. This is given in Equation 4-10. So in Fig. 8-21, if there are, for example, 350 turns on the primary winding and 10 turns on the secondary, the turns ratio is 350:10 = 35:1. The impedance ratio is the turns ratio squared or $35^2:1^2$ = 1225:1. So if R_L is a 4 ohm loudspeaker that is wired across the secondary winding of the transistor, it causes an impedance of 1225 × 4 = 4900 ohms to be reflected into the primary winding. This 4900 ohms is the collector load impedance. The voltage gain of the

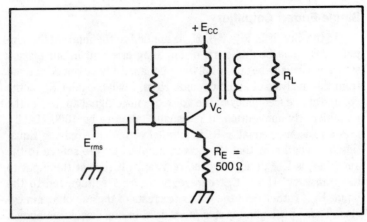

Fig. 8-21. Single-ended power output stage using a transformer.

transistor circuit is 4900 ohms/500 ohms (where R_E = 500 ohms) or 9.8. If beta is equal to 100, the power gain is 9.8 × 100 = 980.

Now let us say that the E_{In} supply furnishes 0.3mW or 0.3 × 10^{-3} watts to the transistor. With a power gain of 980, 0.3 × 10^{-3} × 980 = 0.294 watts is at the output and across the primary winding of the transformer. This power is induced into the secondary winding and fed to the 4-ohm loudspeaker, R_L.

How does power get transferred from one winding of the transformer to the second winding? This is easy to see if you note that in order for 0.294 watts to be developed across the 4900-ohm impedance in the primary, the voltage across the winding, using Equation 3-6, is equal to the square root of the product of the power and impedance or $[(0.294)(4900)]^{1/2}$ = 38 volts rms. If the turns ratio is 35:1, then 1/35 of 38 volts or 1.08 volts rms must be across the secondary. When that voltage is across 4 ohms, $(1.08 \text{ volt})^2/4$ ohms = 0.294 watts, is across the loudspeaker. This is the same power originally applied to the primary.

In this type of circuit, idling dc collector current is set as required for the circuit to deliver the necessary power. Because the dc winding resistance of the primary winding is considered ideally as being zero ohm, the idling collector voltage, V_C, is $+E_{CC}$. There is no dc voltage drop across the ideal zero-ohm transistor winding.

When ac is applied to the input of the transistor circuit, a magnified version of the input signal is developed across the primary of the transformer. This signal varies from a peak of $+E_{CC}$ volts to a crest of $-E_{CC}$ volts. Voltage at the collector is equal to the dc supply voltage, $+E_{CC}$, minus the ac voltage across the transformer at any instant. Thus, when the ac voltage across the primary winding is at its peak, the voltage at the collector, $+E_{CC}$, is equal to the supply voltage less the peak voltage, or zero volt. When the point in the cycle is such that the voltage across the primary is at its crest, $-E_{CC}$ volts, the voltage at the collector is equal to the supply voltage minus the crest voltage, $[E_{CC} - (-E_{CC})]$, or $+2E_{CC}$ volts. Thus, the voltage at the collector can vary from zero volt to $+2E_{CC}$ volts when a transformer is used in the circuit.

In this example, 38 volts rms is across the primary winding of the transformer. Because of this, the peak (or crest) voltage across the winding is 38 volts × 1.414 = 53.7 volts. For the circuit to function, E_{CC} must be at least 53.7 volts. With the applied input signal, the collector voltage will vary from E_{CC} plus the peak voltage during one half of the cycle to E_{CC} minus the peak voltage during the alternate half of the cycle, or from zero to 53.7 + 53.7 = 107.4 volts. Voltage across the transformer will vary in the same way. This voltage variation is illustrated in Fig. 8-22.

Classes of Amplifiers

The operation described here is for a *class-A* amplifier. This means that the transistor conducts through the entire cycle. Here, a sinusoidal input signal was fed to the input of the transistor in

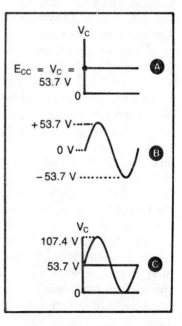

Fig. 8-22. Collector voltage. A. Dc idling voltage is 53.7 volts B. Variations of voltage across the output load or transformer primary. C. The total collector voltage is the sum of the voltages in A and B. As the voltage across the transformer goes through its cycle, collector voltage varies from 0 to 107.4 volts.

Fig. 8-21. The collector voltage and base bias was set so that the transistor is performing as a class-A amplifier to faithfully reproduce the entire ac signal all by itself.

For example, assume that the input signal at the base of the transistor is 2 volts rms. Because the peak voltage is the rms voltage multiplied by 1.414, the ac input signal voltage varies from 2 volts × 1.414 = 2.828 volts at the peak to − 2.828 volts at the crest. The transistor conducts when the applied voltage makes the base positive with respect to the emitter. When the negative crest voltage is applied to the base of the transistor, it may stop conducting. To prevent this from happening, the idling base voltage must be great enough so that the transistor will conduct even when the ac is negative, which is − 2.828 volts in this example. In order to accomplish this, the transistor must idle at more than 2.828 volts dc above the 0.7 volt across the base-emitter junction. Therefore, when idling, the emitter current must be the proper magnitude to develop more than 2.828 volts across the emitter resistor, R_E. By conducting through the complete cycle after these conditions have been satisfied, the transistor is performing as a class-A amplifier.

Class-A amplifiers are used only when low output power is required. When more power is needed, *class B* operation is utilized. Here, two transistors are put into a circuit to reproduce the entire cycle. A circuit for illustrating this type of operation is shown in Fig. 8-23. Two transformers are shown here, but modern circuits normally don't use them.

A transistor is connected across each half of the driver transformer winding. There is no dc supply voltage between the base and emitter of either transistor, so there is no idling current flowing through these devices. Zero collector current is present despite the presence of an E_{CC} supply voltage.

The input signal is applied to the primary of the driver transformer, T_1, and induced into its secondary. There is a center-tap on the secondary so that equal signal voltages are across each half of the winding. The signal voltage from one-half of the winding is applied to the base-emitter circuit of Q_1, while the signal from the second half of the winding is applied to the base-emitter junction of Q_2. One transistor conducts during each half-cycle.

If the half-cycle is such that the upper lead of the driver transformer secondary is positive with respect to the center-tap, the base of Q_1 is positive with respect to its emitter. Voltage across the base-emitter junction is now the proper polarity for that junction to conduct and the transistor is turned on. The amplified version

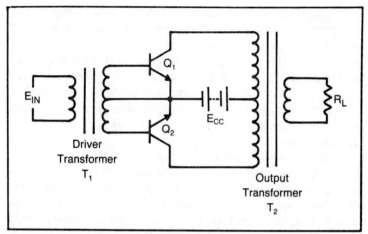

Fig. 8-23. Basic class-B amplifier.

of the half-cycle appears across the upper half of the T_2 winding. At the same time, the bottom lead of the T_1 secondary is negative with respect to the centertap. This makes the base of Q_2 negative with respect to its emitter. Consequently, Q_2 remains off during this half-cycle.

During the second half-cycle when the polarity of the signal is reversed and the bottom lead of T_1 is positive with respect to the center-tap while its top lead is negative with respect to the center-tap, Q_2 conducts and Q_1 remains off. During this half-cycle, the amplified output is developed across the bottom half of the T_2 primary.

The two half-cycles—one-half developed across the upper half of the T_2 primary and one-half developed across its lower half, are both induced into the secondary of T_2. They combine here to form an amplified version of the input signal. The amplified version is applied to load R_L or to a loudspeaker. Because each transistor conducts for only one half-cycle, the total power that the combination can supply is greater than can be supplied by an individual device in a class-A circuit.

You may justifiably ask if the entire cycle is reproduced when you consider that a transistor does not conduct until 0.7 volt is across its base-emitter junction. For example, the transistor will not be turned on and conduct until the input signal level at the base-emitter junction exceeds 0.7 volt. Is it true that the first 0.7 volt of the input signal is lost and only signal voltages above that level are reproduced?

If the circuit in Fig. 8-23 were actually used as shown, this would indeed be the case. But in practical applications, a tiny dc bias current is applied to the transistors' bases to keep them turned on even when the input signal is low. Because of this bias current, both transistors conduct slightly, even in the absence of any input signal. Conduction does not become substantial until the actual ac input signal is applied to the amplifier. Because there is some conduction at all times, the amplifier is referred to as operating in a *class AB* mode.

In other non-audio applications, bias is such that transistors conduct for less than half of a cycle. Here, the applied ac bias voltage may actually be negative. When the input is the right magnitude and polarity to overcome the negative bias voltage by 0.7 volt, the transistor begins to conduct. This is referred to as a *class C* mode of operation.

In a switching type of operation, also not used in audio, the transistor is biased either on or off. If you do not want it to conduct, there must be a negative dc bias voltage between the base and the emitter. In order to conduct, the polarity of the dc voltage applied between the base and emitter must be reversed.

Transistors run hot when they conduct large amounts of current in high-power amplifier circuits. To keep them reasonably cool, they are mounted on metal forms with fins, known as *heat sinks*. These heat sinks conduct heat away from the transistors. Silicon grease is usually used between the transistor and the sink to make the conduction of heat more effective. In some cases, a mica or teflon insulator is used between the transistor and sink to electrically insulate the transistor from the sink. If the insulator is used, silicon grease is also spread over both sides of that insulator to achieve better conduction of heat from the transistor.

Push-Pull Circuits

The class AB audio circuit is known as a *push-pull* amplifier. Only one transistor conducts fully at any time. Several practical circuits involving push-pull arrangements discussed are shown below.

A circuit still being used frequently in portable radios is shown in Fig. 8-24. Here, the input signal is amplified and applied to the primary winding of a driver transformer. As shown in the figure, there are two identical secondary windings on the transformer. Transistors are connected to these windings so that Q_1 conducts during one-half of the ac cycle and Q_2 conducts during the alter-

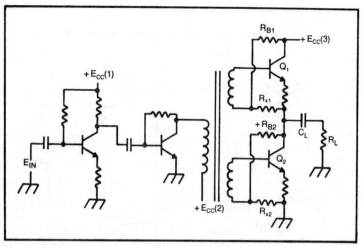

Fig. 8-24. Push-pull circuit using a driver transformer.

nate half-cycle. Note that the bias arrangements involving R_B - R_x combinations at the two transistor base-emitter circuits are identical to those shown in Fig. 8-20D. The difference is that here only one-half of E_{CC} is across each transistor/resistor combination. These resistors, along with $E_{CC}/2$, are used to keep the transistors turned on so that they conduct some slight amount of collector current even when the applied input signal voltage across the base-emitter junctions is less than 0.7 volt. The quantity of collector current conducted varies with the input signal at these low-voltage levels.

Resistor R_L or a loudspeaker is the load. Q_1 conducts the signal to R_L during one half-cycle, and Q_2 conducts it during the alternate half-cycle. The half-cycles combine across R_L to form an amplified version of E_{In}.

A completely transformerless push-pull circuit is shown in Fig. 8-25. Here, the quiescent voltage at the junction of Q_4 and Q_5 is one-half of E_{CC}. This voltage is applied to the base-emitter circuit of Q_1 through the R_{B1} - R_{x1} combination to put Q_1 into a class-A amplifying state. Its collector current flows through R_{x2}, R_{B2}, and R_{B3}. The voltage developed across R_{x2} due to this current is applied between the bases of npn transistor Q_2 and pnp transistor Q_3. Because the upper terminal of R_{x2} is positive with respect to its lower terminal, polarity is such as to keep the base-emitter junctions of both transistors turned on and conducting at all times. This is the same as if the voltage were applied in a similar fashion to

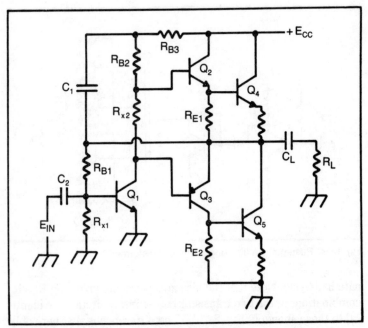

Fig. 8-25. Quasi-complementary push-pull amplifier.

two diodes to keep both of them turned on. But very little current flows through Q_2 and Q_3 when they are idling.

The base currents of Q_2 and Q_3 are amplified by these transistors and appear as collector and emitter currents, respectively, in R_{E1} and R_{E2}. Because of these currents, voltages are developed across R_{E1} and R_{E2}. These voltages turn on Q_4 and Q_5. While the Q_2 - Q_4 combination is a Darlington pair, the Q_3 - Q_5 transistors form a *complementary pair*. It is assigned this name because one device of the pair, Q_3, is a pnp transistor and the second one, Q_5, is an npn transistor. The voltage gain of either pair is 1, while the current power gains are the product of the betas of the two transistors in each pair.

The ac signal voltage is amplified by Q_1 and applied to the Q_2 - Q_3 pair known as *driver* transistors because the output voltage from Q_1 is developed across the R_{B2} and R_{B3} resistors combination. The portion of the signal voltage developed across R_{x2} is negligible because this resistor is very small. As far as the signal voltage is concerned, the bases of both Q_2 and Q_3 are effectively at the same potential. When the signal is such as to make the bases positive with respect to the emitters, the Q_2 - Q_4 combination conducts the ac signal because Q_2 is an npn device. In the alternate half-cycle when the bases are negative with respect to their emitters, the $Q_3 - Q_5$ combination now conducts the ac signal because Q_3 is a pnp device. The half-cycles are recombined across R_L as a magnified version of the input signal.

There are many variations of these circuits. In one, two Darlington pairs are used rather than one Darlington pair and one complementary pair. In another, two complementary pairs are used. In a third, a special circuit is used to stabilize any dc-amplifier, idling-current drift so that an output capacitor, C_L, does not have to be used between the amplifier and load.

Some distortion is present at the output of all amplifiers. To minimize distortion, a parallel RC network is usually wired between the output of the power amplifier and some point at the input voltage amplifier. This RC network establishes a negative feedback loop around the amplifier. How this and other feedback circuits perform their functions will be described in Chapter 10.

Chapter 9

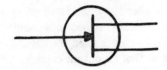

Field-Effect Transistors

Bipolar junction transistors have been depicted as being basically current amplifiers. A specific current is fed to the base circuit. Current equal to beta multiplied by the base current flows in the collector. Special circuits and components involving the transistor are required to make the overall arrangement into a voltage or power amplifier.

Field-effect transistors (FET) can be thought of as exclusively voltage amplifiers. Its input circuit presents a high impedance. A voltage is fed to the high-impedance input circuit. Very little current flows here because of this high impedance, but current flows in the output circuit. The quantity of current in this circuit is dependent upon the voltage applied to the input as well as upon the characteristics of the particular FET being used.

Current in the output circuit usually flows through a resistor, and a voltage is developed across this resistor. The ratio of this voltage to the input voltage is the voltage gain of the FET amplifier stage. Keep in mind that the input current is negligibly small and may be considered nonexistent. Thus, extremely little or no power is needed from the signal source to feed the input transistor circuit. This high-input impedance is a basic characteristic of the FET when used in two of its circuit applications.

There are basically two groups of FETs in terms of their structure and response characteristics. In each group, you will find two types—one which uses a positive power supply when the reference terminal of the transistor is at ground, and one that uses a negative power supply. This indicates that there are two categories of FET devices similar to the npn/pnp types of bipolar junction transistors.

The two groups of transistors are constructed in different ways, but their behavior is similar. In each group, I will describe the type of transistor which uses the positive power supply. These devices are referred to as n-channel FETs. The other type is referred to as a p-channel device. The details supplied for the n-channel transistors also apply to the p-channel FETs. In the latter case, the polarities of the voltage supplies are the reverse of those shown for the n-channel devices.

JUNCTION FETs

The junction FET or JFET is constructed from two semiconductor slabs that are in intimate contact with each other. A drawing of the structure of an n-channel JFET as well as its schematic representation is shown in Fig. 9-1. P-channel devices are constructed in the identical manner, but the types of semiconductor materials are interchanged. As for the schematic representation, the only difference is the directions of the arrows at the gates, which differ by $180°$. The arrow in the p-channel transistor schematic points away from the channel.

Just as was the case with the bipolar junction transistor, these are three terminals that are used. There is one terminal at each end of the channel, known as the *source* and *drain*. A third termi-

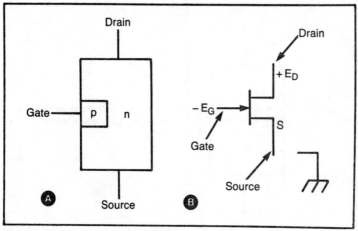

Fig. 9-1. N-channel JFET. A. Structure. B. Schematic symbol.

nal, labelled the *gate*, is at the semiconductor slab adjoining the channel.

The p-slab forms a junction with the n-slab just as it did in the junction diode. In the absence of any applied voltage between the gate and channel, a depletion region exists at the junction of the two materials. This is shown in Fig. 9-2. If the drain is positive with respect to the source, there is plenty of room for electrons to travel in the channel from the source to the drain. This current is close to maximum when the gate is connected to the source because now the depletion region is small. This region is reduced to nothing if the gate is biased somewhat positive with respect to the source. Now electrons can travel more easily from the source to the drain because the available area that is free of the depletion region has been increased to a maximum.

The depletion region is increased if the gate is reverse biased with respect to the source. Because of this, less space is available in the channel for the electrons to flow. As the reverse voltage is increased, the number of electrons that can flow in the channel is reduced further. It drops to just about zero current when the reverse bias voltage is very high. This is referred to as the *pinch-off* voltage.

Because of this depletion region, the drain current is dependent upon the gate-to-source voltage. The ratio of the drain current to the negative gate voltage is an important characteristic of all types of FETs.

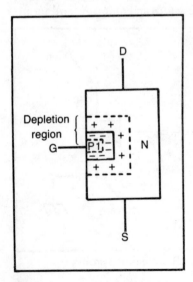

Fig. 9-2. Depletion region exists between the gate and channel of the JFET.

IGFETS AND MOSFETS

The abbreviation IGFET stands for *insulated gate FET*, while the abbreviation MOSFET stands for *metal oxide semiconductor FET*. These are just two names for identical devices.

Two categories of transistors fall into the MOSFET group. One category, the *depletion type*, is very similar in performance to the JFET. The amount of current that can flow between its source and drain decreases as the voltage between the gate and source is made more negative and increases as the gate is made more positive with respect to the source. But this is where the two types differ. You do not want, ideally, any conduction between the gate and channel. The JFET will conduct if the voltage at the gate is more than about + 0.5 volt higher than the source. Therefore, the gate of the JFET can be made quite negative with respect to the source, but its positive swing is limited to about + 0.5 volt. These negative gate-to-source voltage characteristics also apply to the MOSFET. However, the gate of this type of transistor *can* be made quite positive with respect to the source. Because of this, the source-to-drain current increases when the applied gate-to-source voltage increases above 0 volt. Here the + 0.5 volt limit does not exist. This is because the MOSFET is not a junction device.

The structure of a depletion type of MOSFET along with its schematic representation is shown in Fig. 9-3. This drawing is for the n-channel device. Opposite types of semiconductor materials are used to form p-channel devices. The schematic of the p-channel devices show the arrow at the substrate drawn in the opposite direction to that shown in the diagram.

The transistor is built into a p-slab foundation known as the *substrate*. (An n-slab is used as the substrate for a p-channel device.) An n-slab is diffused onto the substrate. As in the case of the JFET, one end of the n-slab is the source and the second end is the drain. The gate is simply a thin layer of conducting metal that is separated from the n-slab by an oxide. This oxide insulates the gate from the n-slab.

When the gate is connected to the source so that the gate is at 0 volt with respect to the source, a specific current flows from the source to the drain. This current depends upon the transistor being used. Should the gate be made negative with respect to the source, an electric field is formed in the n-slab. The conducting area of the slab is effectively narrowed to restrict the flow of electrons. Similarly, a positive voltage at the gate with respect to the source causes the field that existed when the gate-to-source voltage was

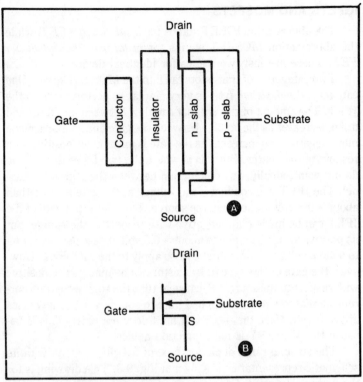

Fig. 9-3. N-channel depletion-type MOSFET. A. Structure. B. Schematic representation.

zero to be reduced. Now more current can flow through the n-channel.

A diagram of the internal structure of an n-channel *enhancement type* MOSFET is shown in Fig. 9-4. This is the second category of MOSFET. Here, two individual n-type semiconductors are different into the substrate. There semiconductors do not touch each other. Because of this, there is no conduction between the source and drain when the gate-to-source voltage is zero or negative. Conduction starts when a positive voltage is applied to the gate with respect to the source. Drain current increases as this voltage is made more positive. This is because a conducting path is now established in the p-substrate.

A capacitance is actually produced by the conducting material comprising the gate and the substrate. When the gate is made positive and the p-material forming the substrate is relatively negative, there is a surplus of electrons in the substrate. These link the source to the drain to establish a path in which the electrons can flow. This surplus of electrons increases as the gate is made more positive, so conduction from the source to the drain also increases.

The substrate lead in the schematic is shown as unattached. When MOSFETs are used, the lead from the substrate is usually connected to the source. It seldom gets any more attention than that in the description of a circuit.

CHARACTERISTICS OF FETS

Bipolar transistors have been grouped by their alphas and betas. FETs are characterized by their *transconductance* or g_m. The transconductance tells you how much the drain current will change in comparison with a change in the gate-to-source voltage. The mathematical symbol for change is *delta,* Δ. Using this notation, the formula for the transconductance of a specific FET is

$$g_m \quad \frac{\Delta I_D}{\Delta V_{GS}} \qquad \text{Eq. 9-1}$$

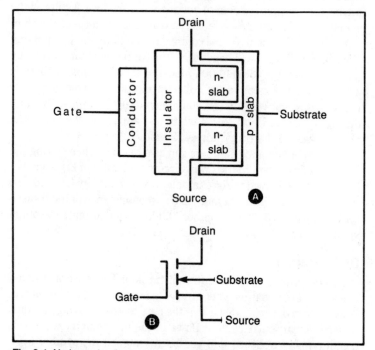

Fig. 9-4. N-channel enhancement-type MOSFET. A. Structure. B. Schematic representation.

where ΔI_D is a specific change in drain current for ΔV_{GS}, a particular change in gate-to-source voltage.

Looking at the formula, you will note that transconductance is a current divided by a voltage. In Equation 2-2, you saw that resistance is voltage divided by current. So if resistance is a factor indicating to what degree a component resists the flow of electrons, conductance (or transconductance) being just the opposite or resistance, indicates how well a component conducts electrons. Thus, a better-conducting material has a lower resistance and a higher conductance of transconductance number. Because one item is the opposite of the other, resistance is specified in *ohms* and conductance in *mhos*. Mhos is the unit of resistance, ohms, spelled backwards. Lately, conductance has been assigned the name *Siemens*, supposedly to replace the term mho. But the mho is still the more popular unit. The transconductance of an FET is specified in micromhos (1/1,000,000 of a mho) and in millimhos (1/1000 of a mho).

For example, assume you are feeding an ac signal with a 1.2 peak-to-peak voltage to the gate-source junction of an FET. This means that the gate-to-source voltage will vary by 1.2 volts as the signal goes through its cycle. If the transconductance of the transistor being used is 6000 micromhos or 0.006 mhos, by applying Equation 9-1 you will find that the drain current varies by $(\Delta V_{GS})(g_m) = (1.2 \text{ volts}) (0.006 \text{ mhos}) = 0.0072$ amps. If there is a 10,000-ohm resistor in the collector circuit, the voltage across that resistor will vary by $(10,000 \text{ ohms}) (0.0072 \text{ amp}) = 72$ volts during the cycle. Because 1.2 watts was applied to the input, the voltage gain of the transistor is 72 volts/1.2 volts = 60.

Transconductance numbers can also be used when setting the quiescent dc current in the drain circuit. The ratio in Equation 9-1 holds fairly well for dc. For example, if you want 1 mA (or 10^{-3}A) to flow in the drain circuit and the transconductance of the transistor being used is 5000 micromhos (0.005 mhos), the gate-to-source voltage must be set at $I_D/g_m = 10^{-3}$A/0.005 mho = 0.2-volt.

Drain Current

In describing how the different types of FETs work, I noted one important situation where the gate is connected to the source. Another way of saying this is that the gate-to-source voltage is equal to zero. A particular current flows in the drain circuit when this condition is established and has been assigned the symbol I_{DSS}.

I_{DSS} is the maximum drain current that flows in the JFET without making the gate positive with respect to the source. In the enhancement type of MOSFET, this current is just about equal to zero. Here the gate must be made positive with respect to the source if any drain current is to flow. As for the depletion type of MOSFET, I_{DSS} is similar in magnitude to the I_{DSS} of the JFET. When the gate of any type of MOSFET is made positive with respect to the source. Under this condition, more drain current than I_{DDS} can flow in either type of MOSFET. This positive bias voltage is undesirable for the JFET. Here, this voltage should never exceed $+0.5$ volt. The importance of observing this limit cannot be overemphasized.

Leakage Current

I_{BO} was the symbol used for the collector-to-base leakage current of the bipolar transistor. A similar leakage current may flow through the gate-to-channel terminals of an FET. The symbol for this current is I_{GSS}. It is the current that flows in the gate lead when the source is connected to the drain and is extremely small compared to I_{BO}. If a resistor is connected between the gate and source, the polarity of the voltage developed across this resistor is such as to make the gate of an n-channel device positive with respect to its source. But because the leakage current is very small, the gate resistor must be quite large, usually well over 1 megohm, before the voltage developed across can affect the dc bias status of the transistor.

More Characteristics

The following characteristics of the FET discussed here are numerous but are not, in most instances, as important as those described above.

First, there is a *pinch-off* voltage, V_P. This is the negative voltage that must be applied to the gate with respect to the source to reduce the drain current to an infinitesimally small level. V_P is a symbol applied to the JFET. The *threshold* gate-to-source voltage using the symbol V_T or $V_{GS(th)}$ is a term applied to the enhancement type of MOSFET. With V_P or V_T between the gate and source, the drain current is about 10 μA in either device. A similar low drain current level is used for the enhancement/depletion type of transistor to indicate the gate-to-source voltage required for a negligible drain current to flow. Here, the symbol $V_{GS(off)}$ is used for this purpose.

The transconductance also varies with the drain current and

was defined in Equation 9-1. The symbol g_{MO} is used to indicate the transconductance of an FET when the drain current is at I_{DSS}. This is the figure usually specified for transconductance in data sheets supplied by the manufacturer. If this transconductance is known, the transconductance at the actual gate-to-source bias voltage, V_{GS}, used in a circuit, can be derived through use of Equation 9-1.

$$g_m = g_{MO} \left(1 - \frac{V_{GS}}{V_P} \right) \qquad \text{Eq.9-2}$$

There are two regions of operation for FETs. When the drain-to-source voltage, V_{DS}, is less than V_P, the drain-to-source resistance of the JFET channel is also low. If it is operating in the range where V_{DS}, is equal to or less than V_P, the transistor is said to be operating in the *ohmic* or *on* region. When working in this region, the resistance of the channel does not depend only upon the voltage between the drain and source. It is also related to the gate-to-source voltage, V_{GS}, that is applied to the JFET. The transistor can therefore function as a variable resistor in a circuit. Its resistance can be altered by simply resetting V_{GS}.

The transistor can be used as an amplifier with a minimal amount of distortion if its drain-to-source voltage is greater than the pinch-off voltage. Now the impedance or ac resistance between the drain and source is quite high. When referring to operation in this area of the characteristics, the JFET is said to be operating in the *pinch-off* or *off-region*.

FET CIRCUITS

Earlier I described three basic circuit arrangements for the bipolar transistors. FETS similarly can be used in three different basic configurations. These circuits are used primarily to provide voltage gain because the current and power gains are essentially negligible. The circuits are shown as using JFETs. MOSFETs could just as easily have been substituted for the JFETs. Either way, performance will be similar.

Common-Gate Circuit

The FET equivalent of the common-base circuit is shown in Fig. 9-5. All voltages here are referred to the gate. Note that in

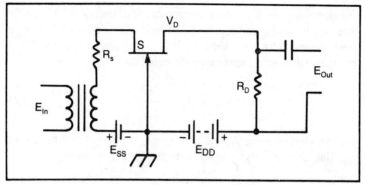

Fig. 9-5. Common-gate circuit.

this as well as in most transistor schematics, an S is placed near one terminal to indicate that it is the source. The gate terminal has an arrow and the drain is the only remaining terminal of the transistor.

E_{SS} in the circuit supplies a fixed gate-to-source bias voltage. It is not the only element in the circuit. When the idling current, I_D, flows between the drain and source, it also flows through R_D and R_S. Dc voltage is developed across R_D. This voltage is equal to $I_D R_D$. Similarly, a dc voltage is developed across R_S and this voltage is equal to $I_D R_S$. $I_D R_S$ must be added to E_{SS} to determine the gate-to-source voltage applied to the transistor. To keep the effect of the voltage developed across R_S at a minimum, it should be a low-resistance component.

The goal of the bias circuit is to establish a desirable idling current in the source and drain circuits of the transistor. E_{DD} is the supply voltage for the drain circuit. As in the case of the voltage at the collector with respect to the supply voltage in a bipolar junction transistor, the dc or idling voltage at the drain of the FET, V_D, should be about one-half of E_{DD}. This condition is established by selecting a suitable drain resistor, R_D, for use in the circuit and then setting the gate-to-source bias voltage at the potential that will cause V_D to be equal to $E_{DD}/2$.

In Fig. 9-5, the input signal is shown being applied through a transformer. This transformer is in series with E_{SS} and R_S. Assuming the transformer has a 1:1 turn ratio, E_{IN} is at both the input and at the output of the transformer. The dc resistance of a transformer is almost always low in comparison with R_S, so the voltage developed across it does not usually affect the gate-to-source bias voltage to any degree worth noting.

Ac signal source E_{IN} sees a resistance, R_{IN}, that is related to the magnitudes of R_S and g_M. As you know from Equation 9-2, g_M is the transconductance of the transistor at its point of operation. The input resistance is approximately equal to

$$R_{IN} = R_S + \frac{1}{g_M} \qquad \text{Eq.9-3}$$

The ac voltage gain, A_V, of the circuit, is also related to R_S and g_M. But now, R_D also comes into the picture, making

$$A_V = \frac{g_M R_D}{1 + g_M R_S} \qquad \text{Eq.9-4}$$

If R_S is small or equal to zero, the voltage gain is simply $g_M R_D$. The voltage gain determined from this equation applies at all frequencies despite the fact that the common-gate circuit is used primarily in high-frequency applications.

As an example of an application of using this arrangement, assume you have a transistor where I_{DSS} = 12 mA, V_P = 6 volts and g_{MO} = 4000 micromhos. The drain supply, E_{DD}, is 18 volts and the source supply voltage is E_{SS} = 1.5 volt. Resistor R_D = 3000 ohms and R_S = 400 ohms. Assume E_{IN} across both windings of the transformer is 1 volt and that the ac and dc resistance of the transformer is 100 ohms. (This transformer resistance is high, but is used here to illustrate its effect on the gain.) This is all shown in Fig. 9-6. What is the idling current, drain voltage, gain, and other characteristics of the circuit?

The desired goal is to have a quiescent drain voltage that is equal to one-half of E_{DD} or 9 volts. To achieve this, 9 volts dc must be across R_D so when this 9 volts is subtracted from E_{DD}, only 9 volts is left at the drain. This is accomplished if I_D is equal to 9 volts/3000 ohms or about 3×10^{-3} amperes. This 3×10^{-3}A also flows through R_S and the secondary of the transformer. The total resistance of the two components is 400 ohms + 100 ohms = 500 ohms. Therefore, a total voltage equal to $I_D R_S$ = (3×10^{-3}A) (500 ohms) or about 1.5 volts, is across the two resistances. When added to the 1.5 volts across the E_{SS} supply, −3 volts is between the gate and source.

The 3×10^{-3}A current is desirable in the drain circuit. What the current actually is can be determined from an equation relation the drain current, I_D, to I_{DSS}, the gate-to-source voltage, V_{GS},

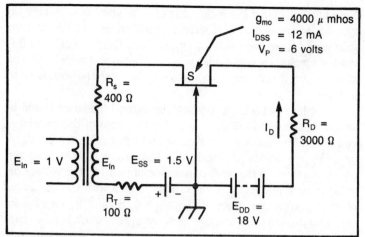

Fig. 9-6. Example used to determine characteristics of common-gate circuit.

and the pinch-off voltage, V_P:

$$I_D = I_{DSS} \left(1 - \frac{V_{GS}}{V_P} \right)^2 \qquad \text{Eq. 9-5}$$

By substituting the transistor data into the equation, you will find that $I_D = [12 \times 10^{-3}\text{A}] [1 - (3 \text{ volts})/(6 \text{ volts})]^2 = 3 \times 10^{-3}\text{A}$. Thus the desired quiescent drain current flows when the total resistance in the source circuit, equal to the sum of R_S and the transformer winding, is 500 ohms.

The transconductance of the transistor when the gate-to-source voltage is zero was specified as being 4000 micromhos. When 3 volts is between the gate and source, the actual transconductance is

$$g_M = 4000 \; \mu\text{mhos} \left(1 - \frac{3\text{-volts}}{6\text{-volts}} \right) = 2000 \; \mu\text{mhos}$$

This is determined using equation 9-2. Therefore the g_M for the transistor under actual operating conditions is 2000 micromhos.

Using equation 9-4, we see that the voltage gain of the transistor circuit is

$$A_V = \frac{(2000 \times 10^{-6}\text{mhos})(3000 \text{ ohms})}{1 + (2000 \times 10^{-6} \text{ mhos})(500 \text{ ohms})} = \frac{6}{2} = 3$$

263

Note how important the magnitude of R_S is when determining the gain of the transistor. If it were a short circuit or equal to zero, gain would have been equal to $g_M R_D = 6$. Output would be 6 (1 volt) = 6 volts. By using this circuit as shown and with 1-volt rms at the input, 3 volts must be across R_D because the gain is equal to three.

In order to check our results, the natural tendency would be to use Equation 9-1. With 1-volt ac at the input of the transistor, you would find that the ac collector current would then be $\Delta I_D = (g_M)(\Delta V_{GS}) = (200 \times 10^{1-6}\text{-mhos})$ (1 volt) = 2 mA. When this current flows through the 3000-ohm collector resistor, the voltage developed here is $I_D R_D = 6$ volts. This agrees with the calculated result shown above when R_S equalled 0. Should R_S differ from zero as it does in this example, the gain of the circuit is considerably less than our calculation would show when using Equation 9-1. There is negative feedback because of the presence of R_S is not taken into account by the equation.

Note that the input impedance of this circuit, as well as of any common gate circuit, is low. Using Equation 9-3, we find it equal to

$$R_S + 1/g_M = 400 \text{ ohms} + 1/2000 \times 1 \text{ V}^{-6} \text{ mhos}$$
$$= 400 \text{ ohms} + 500 \text{ ohms}$$

or a total of 900 ohms. Because of this low impedance, there is a loss of input voltage when a high-impedance source feeds the circuit. For example, if a 1-volt, 100-ohm source were feeding the 900-ohm load, the voltage across the 900 ohms would be

$$\frac{900 \text{ ohms}}{(100 + 900) \text{ ohms}} 1 \text{ volt} = 0.9 \text{ volt}$$

This is derived by applying the voltage divider equation to the circuit in Fig. 9-7. Thus with E_{IN} at 1 volt, only 0.9 volt would be at the input of the transistor. With a transistor gain of 3, 2.7 volts would be across R_D. As a result, the gain of the overall circuit is 2.7 volts/1 volt = 2.7 rather than 3. Gain is less than what it is when the input resistance of the transistor is high. A high-input resistance or impedance is achieved when the transistor is used in a common-source or common-drain circuit.

Common-Source Circuit

The circuit in Fig. 9-8 is probably the most frequently used cir-

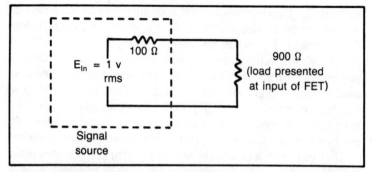

Fig. 9-7. Voltage divider formed by 1-volt, 100-ohm source feeding a 900-ohm load presented by the input circuit of the FET.

cuit using an FET. In the *common-source* circuit, a signal is applied between the gate and ground. Output voltage is developed across R_D. As is always the case, the source resistor, R_S, affects the gain of the stage. In this circuit, the voltage gain is very close to that of the common-gate amplifier and can be approximated by using Equation 9-4. Because a load resistor, R_L is across R_D in this circuit, the resistance of this parallel combination should be substituted for R_D in the equation. The resistance seen by E_{IN} is much higher than what it saw at the input of the common-gate circuit. Here, it is just about equal to R_G. As for the output impedance seen by load resistor R_L it is slightly less than R_D.

The dc voltage between the gate and source sets the quiescent idling conditions for the transistor. Resistors R_G and R_S are between these two terminals of the FET. Only leakage current flows

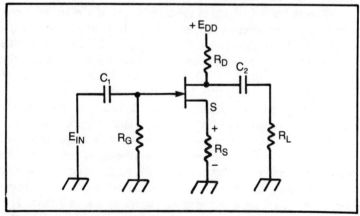

Fig. 9-8. Common-source circuit.

through R_G. Because this current is usually very small, a negligible voltage is developed across R_G, but drain or source current flows through R_S. When the current flows, a voltage with the polarity shown is developed across the resistor, making the gate negative with respect to the source. The transistor is reverse biased by this voltage. Voltage across R_S is equal to the product of the drain current, I_D, with R_S. R_S is chosen so that this product with I_D, or the voltage developed across it, is the proper bias voltage to establish the desired operating conditions for the transistor.

For example, let's use the transistor from the previous problem. Here, g_{MO} = 4000 micromhos, I_{DSS} = 12 mA and V_P = 6 volts. Setting E_{DD} equal to 21 volts, you want a total of 1/2 of 21 volts or 10.5 volts, to be developed across R_D and R_S when the circuit is idling. If R_D is chosen as 2500 ohms and R_S is 1000 ohms, the total resistance in this circuit is 3500 ohms. When 10.5 volts is across the transistor, 10.5 volts/3500 ohms = 3×10^{-3}A flows in these resistors. Because R_S is 1000 ohms, 3×10^{-3}A) (1000 ohms) = 3 volts is developed across the resistor. This is the bias voltage between the gate and the source. It was the same bias voltage used previously in the problem involving the common-gate circuit. There it was found that at this gate-to-source voltage and with the particular transistor being used, the G_M at the point of operation is 2000 micromhos. Using all this information and Equation 9-4, you will find that the ac voltage gain of the transistor circuit is $g_M R_D/(1 + g_M R_S)$ = 2000 $\times 10^{-6}$ mhos) (2500 ohms)/[1 + (2000 $\times 10^{-6}$ mhos) (1000 ohms)] = 1.67. The input impedance has no effect on voltage gain because the resistance of R_G is usually much larger than the resistance of the ac signal source, E_{IN}.

R_L was not considered in the calculations. If it is 10,000 ohms, the equivalent resistance formed by R_L with R_D using Equation 3-2 is (10,000 ohms) (2500 ohms)/ (10,000 ohms + 2500 ohms) = 2000 ohms. Applying this information to Equation 9-4, the gain of the overall circuit is $(2 \times 10^{-3}$-mhos) (2000 ohms)/[1 + (2 $\times 10^{-3}$-mhos) (1000 ohms)] = 1.33, rather than 1.67.

Thus, gain is low. This is because R_S had to be very large to establish the desired dc bias voltage. If a large capacitor were connected across R_S, it would not upset the dc voltage developed across it. V_{GS} and drain current would remain as before, but the capacitor would be a short circuit for any ac signal current flowing in that resistor. In fact, as far as the signal is concerned, the resistance between the source and ground would appear as 0 ohms. This assumes that the shunting capacitor is large. Equation 9-4 can still be used to determine the ac gain of the circuit, but because the capacitor shunts R_S, R_S in the equation would be set equal to 0 ohms. Now the gain of the circuit is $(2 \times 10^{-3}$ mhos$)$ $(2000$-ohms$)/[1 + (2 \times 10^{-3}$ mhos$) (0$ ohms$)] = 4$. This is quite an improvement over the gain of 1.33 when no capacitor was used to shunt R_S.

Even a gain of four is quite low. To see how to improve on this, you must analyze Equation 9-4. When $R_S = 0$, gain is $g_M R_D$. Noting this, it is obvious that the gain of a circuit will be much higher if the resistance of R_D and the $R_D = R_L$ combination are increased or if g_M is high. The latter goal is achieved when the transistor is biased so that the gate-to-source voltage is as close to zero as possible. Of course, the gain would be much higher if the g_{MO} of the transistor used in the circuit were higher. A circuit using a transistor with a high g_{MO} that is biased with a low V_{GS} voltage and loaded with a large resistor at the output will produce the needed high gain.

Common-Drain Circuit

The FET equivalent of the common-collector circuit is shown in Fig. 9-9. While the voltage gain in this common-drain circuit is just slightly below 1 as it was for the emitter follower, the input impedance is R_G and the output impedance is just slightly lower than $1/g_M$.

For example, let $g_M = 2000$ micromhos, $R_S = 3500$ ohms, $R_G = 0.5$ megohms, and $R_L = 10,000$ ohms. While the input impedance is R_G or 0.5 megohms, the output impedance seen by R_L is $1/g_M = 1/2 \times 10^{-3}$ mhos $= 500$ ohms. The output voltage is

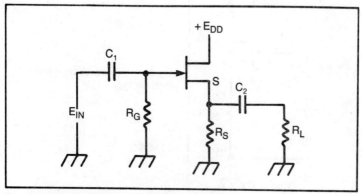

Fig. 9-9. Common-drain (source-follower) circuit.

just slightly less than E_{IN} because the voltage gain is just below 1. If R_L = 10,000 ohms, it has little effect on the output voltage because it is much larger than the output impedance of the transistor.

BIAS CIRCUITS

Different circuits are used to establish the gate-to-source dc bias voltage of the field-effect transistors. A method using a battery, E_{SS}, was illustrated in Fig. 9-5. In Fig. 9-8, the entire dc bias voltage was developed across R_S and no battery was used. In some circuits, it is desirable that R_S be very large. Although this reduces ac gain of the circuit, it also reduces any distortion of the signal due to the transistor circuit. But if R_S is made very large, a large negative voltage will be established between the gate and source, putting the transistor near pinch-off. In order to keep this from happening, the circuit in Fig. 9-10 is used.

In the figure, R_S is assumed to be high enough so that a large voltage is developed across it. E_{GG} is added to the circuit to counter the voltage developed across R_S. E_{GG} is of such magnitude and polarity that when it is subtracted from the voltage across R_S, a negative voltage still remains at the gate of the transistor with respect to its source. Now the voltage at the gate is less negative than it would have been in the absence of E_{GG}. The transistor is set to operate with a relatively high g_M because of this less negative (more positive) gate voltage with respect to the source.

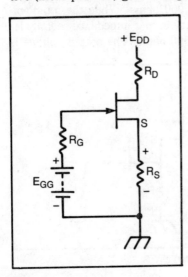

Fig. 9-10. Bias circuit used when R_S must be very large.

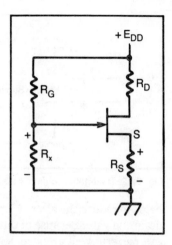

Fig. 9-11. Frequently used bias circuit.

A frequently used bias circuit is shown in Fig. 9-11. Here, no battery is used between the gate and R_S. A voltage is developed across R_x due to the voltage-divider action of resistors R_G and R_x. The voltage across R_x counters that developed across R_S. The difference of the two voltages establishes the desired gate-to-source voltage without the use of a battery. In fact, the voltage across R_x is the same as would be supplied by E_{GG} in Fig. 9-10, but now there is no battery in the circuit. The only drawback here is that the ac input impedance is the parallel combination of R_G and R_x, so the input impedance is less than it would have been had only R_G been in the circuit. Despite this drawback—and it is a relatively minor one—this bias circuit is used in much of the available equipment built around FETs.

JFETS are shown here as well as in the various circuit arrangements. This information and circuitry also apply to MOSFETs.

DUAL-GATE FETS

The transistors shown above had only one gate. Most FETs do have only one gate, but there are also types with two gates. The schematic representations of these devices are shown in Fig. 9-12. Here, the n-channel versions are shown. The only difference between the n-channel and p-channel transistors is in the direction of the arrows.

Suppose that a signal is applied to either device between gate 1 and ground. Both gates are biased to establish the device's drain current. The gate without the applied signal is frequently by-passed to ground by a capacitor. The added capacitor at the second gate,

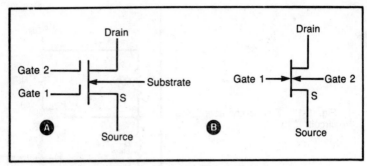

Fig. 9-12. Dual-gate n-channel transistors. A. MOSFET. B. JFET.

helps to reduce the already small capacity that exists between the drain and gate. Because of the low internal capacity, the transistor can perform well at extremely high frequencies.

When experimenting with this type of transistor, you should note two important facts:

1. Transconductance of the transistor drops when the gate-to-source bias voltage used around the device is reduced.

2. Drain current is also related to the gate #2-to-source voltage. This current also drops as that voltage is reduced.

Both gates affect drain current. Because of this, if two different signals are applied to the two gates, a combination of the two appears at the output. Dual-gate FETs are therefore useful in switching and radio-frequency modulation applications as well as for use at extremely high frequencies.

POWER FETS

New variations on the junction FET are constantly being developed. The types discussed above generally apply to only small signal devices. But newer types of FETs are available for use as power amplifiers to drive transducers. A device that converts electrical into mechanical energy, such as loudspeakers, is known as a transducer. Among the devices used for this purpose, you will find the VFET. A HEXFETs is the MOSFET version of the VFET. A typical output stage from an amplifier using VFETs is shown in Fig. 9-13. Here Q_1 and Q_2 are both VFETs.

I described push-pull operation in the discussion of bipolar transistors. A push-pull signal is applied to the circuit in Fig. 9-13. Signal voltages for application here may be generated in circuits using bipolar devices as described in the previous chapter.

270

Q_1 in Fig. 9-13 is biased by the $R_{G1}/R_{x1}/R_{S1}$ circuit, while Q_2 is biased by the R_{G2}/R_{x2} circuit. These bias circuits are like the one shown in Fig. 9-11. One-half of the sinusoidal input signal is applied to Q_1 and the second half is applied to Q_2. During the positive half-cycle, the gate of Q_1 is positive with respect to ground, so it conducts. During the negative half-cycle, the gate of Q_2 is positive with respect to the $-E_{DD}$ supply, so now it conducts. The two halves of the signal are recombined across R_L.

In order to keep distortion low, the dc bias is set so that the idling drain currents of Q_1 and Q_2 are large. However, transistors get hot due to the power dissipated when drain current flows. But drain current does not increase as the temperature of the FET rises. Here, additional drain current is added to the circuit to push the temperature of the transistors above the maximum limit, so the transistors will not overheat and be destroyed. This overheating can occur in bipolar-junction power transistors because current in these devices does tend to increase with the temperature of the junctions.

TRANSISTOR CIRCUITS

Except for the push-pull circuit in Fig. 9-13, one transistor was

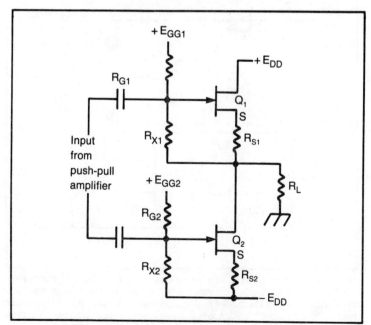

Fig. 9-13. VFET power amplifier.

shown in each example when I discussed the various circuit designs. Actually, circuits may consist of a number of FETs or combinations of FETs with bipolar-junction devices. Several typical circuit arrangements are shown in Fig. 9-14.

In Fig. 9-14A, Q_1 is in an ordinary common-source circuit. Its input impedance is the resistance of the parallel combination of R_{G1} and R_{x1}. Q_1 amplifies E_{in}. The amplified signal appears at the gate of source-follower Q_2. The gain of the source follower is just below one, but its output impedance is low. This may be an important requirement.

Many different combinations of FETs are found in electronic circuits. They may be combined as shown in Fig. 9-14A or may simply be in a sequence of common-source circuits.

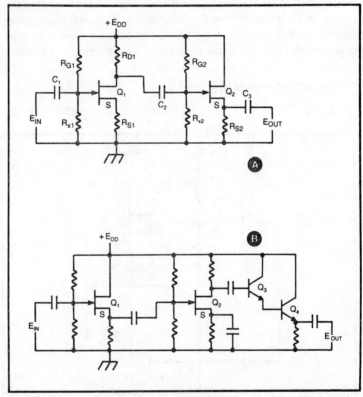

Fig. 9-14. FET circuits. A. Circuit using only FETs presents a high input impedance and a low output impedance. B.Circuit with a very high input impedance and a very low output impedance using a combination of transistor types. The low output impedance is achieved using a Darlington pair. A source follower is used at the input to present the high input impedance.

Another configuration using two JFETs, but this time with two bipolar devices added to the circuit is shown in Fig. 9-14A. The input impedance is very high because a source follower is used for Q_1. High amplification is achieved through use of a JFET circuit with the emitter resistor by-passed by a capacitor. This is Q_2 in the circuit. The output impedance is very low because a Darlington pair is used to supply the output voltage, E_{out}, to a load.

MOSFETs can be used in similar circuits. The input impedance presented by these devices is usually much higher than that provided by the JFET. However, caution must be used when handling a MOSFET. If you rub the leads with your fingers, a static charge will develop on the elements of the device. This charge may be of sufficient voltage to cause breakdown between the various sections of the transistor.

When using a MOSFET, keep all leads together until you are ready to wire the transistor into the circuit. It is best to ground one of your wrists before separating the leads. This can be done by simply connecting a clip lead from a water pipe or radiator to a metal watch band. Even so, you should not touch the leads any more than is necessary.

Chapter 10

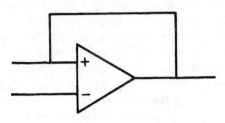

Feedback

When common-emitter transistor circuits were described, I noted that gain is reduced when there is a resistor between the emitter and ground. A similar relative gain situation exists in the common-source FET circuit when a resistor is between the source and ground. In the phonograph playback amplifier in Fig. 8-19B, the feedback principle was used to generate the frequency characteristic that was needed. To improve performance it was also suggested that signal be fed back from the output of a power amplifier to its input.

All of the above situations are commonly used in electronics and are referred to as *feedback circuits*. The signal is fed back in some way from the output of the amplifier, then combined with the signal fed to the input of the overall circuit. The combination of the two signals is reapplied to the input circuit.

Signal fed back from the output of an amplifier can be in phase with the ac voltage at the input. This means that when one signal is added to the other, the total voltage is greater than the input signal voltage by itself. This addition of two ac sinusoidal signals was illustrated in Chapter 4. Because the total signal at the input is now greater than the applied input signal by itself, and now total is applied to the input, the signal at the output of the amplifier is now greater than what it would have been had only the original input signal been present. This is known as *positive feedback*.

Should the feedback signal be 180° out of phase with the ac

input voltage, the total of the two signals is less than the input voltage by itself. If the difference is fed to the input, the output would be at a lower voltage than the original input voltage by itself. This is referred to as a *negative feedback* circuit.

Both types of feedback circuits are very useful. Despite the lower gain of the negative feedback arrangement, it applications help improve the performance of many circuits. Positive feedback may cause problems in different circuits, but its applications in generating audio-and radio-frequency signals are very common.

GAIN WITH FEEDBACK

Amplifier voltage gain, A_v, without feedback, was detailed in the previous two chapters. The ac output voltage, E_{Out}, is simply equal to the input signal voltage, E_{In}, multiplied by that gain. But now let's take a portion of the output signal and feed it back to the input. This can be accomplished as shown in Fig. 10-1.

A straight amplifier is in Fig. 10-1A. Here E_{In} is the input signal and E_{Out} is the output signal. The ratio of E_{Out} to E_{In} is the gain of the circuit. No feedback is used. But note that E_{Out} is 180° out of phase with E_{In}. That means that when the upper terminal of E_{In} is in the portion of the cycle that is positive with respect to ground, the upper terminal of E_{Out} (at the collector of the transistor) is negative with respect to ground. The reverse situation exists in the alternate half of the cycle.

In Fig. 10-1B, the output voltage is divided between resistors R_1 and R_2. Using the voltage divider equation, the voltage across R_2 is $E_{Out} [R_2/(R_1 + R_2)]$. The voltage across R_2 is fed back to the input and placed in series with E_{In}. This is shown more clearly in Fig. 10-1C where the connections to R_2 are the same as before, but the resistor is relocated to show how the voltage across R_2 is in series with E_{In}. Note the 180° phase difference of the two voltages.

In A of the figure, the entire input voltage, E_{In}, was fed to the transistor. In C, the sum of E_{In} and the voltage developed across R_2 is fed to the input of the transistor. Because the two voltages are out of phase, the sum of these voltages is less than E_{In}. So a voltage that is less than E_{In} is fed to the base-emitter circuit in this feedback arrangement. The transistor has the same gain here as in A, but now the input signal is lower than before. Because the transistor circuit gains are identical in both instances, the output voltage is lower in the circuit shown in C than it was in the circuit shown in A of the figure. This reduction in overall gain is actually a result of having an out-of-phase signal fed back from the output to the input. Thus, there is *inverse* or negative feedback in the circuit.

The gain of a feedback circuit can be calculated using Equation 10-1.

$$A_f = \frac{A_v}{1 - BA_V} \qquad \textbf{Eq. 10-1}$$

Here, A_v is the gain of the circuit before feedback is applied. For example, this is the voltage gain of the circuit in Fig. 10-1A. In the equation, B represents the fraction of the voltage at the output that is fed back to the input. In our example, $B = R_2/(R_1 + R_2)$. This ratio is derived from the voltage-divider equation.

For example, let us assume that the voltage gain, A_V, of the circuit in Fig. 10-1A is 60. Fig. 10 B and C, let $R_1 = 116,000$ ohms and $R_2 = 4000$ ohms. Then $B = 4000$ ohms/(116,000 ohms + 4000 ohms) = 1/30. BA_V for the equation is (1/30)(60) = 2. Before substituting this into the equation, you must note if feedback is positive or negative. If it is negative, $-BA_V$ must be substituted into the equation and if it is positive, $+BA_V$ must be put into the equation. In the circuit in Fig. 10-1B, feedback is negative. Therefore -2 must be substituted for BA_V.

$$A_f \quad \frac{60}{1 - (-2)} = \frac{60}{3} = 20$$

With feedback, the gain of the overall circuit has dropped from 60 to 20. So if $E_{In} = 0.1$ volt, the output from the circuit in Fig. 10-1A is $0.1 \times 60 = 6$ volts. The output from the circuits in Figs. 10-1B and 10-1C is $0.1 - 20 = 2$ volts. Here you have a 66.7% loss of output voltage because of the inverse feedback.

The advantages of inverse feedback are numerous. All of these can be calculated by using the *feedback factor*, the denominator in Equation 10-1. In our example, the feedback factor is 3.

It is frequently desirable that the input impedance of a circuit be high. This is the impedance seen by E_{In}. Inverse feedback increases the input impedance present without feedback by the feedback factor. This means that is is equal to the input impedance without feedback multiplied by the feedback factor. If you use 3 as the feedback factor, and the input impedance of the circuit is 10,000 ohms without feedback, it is increased to $3 \times 10,000$ ohms

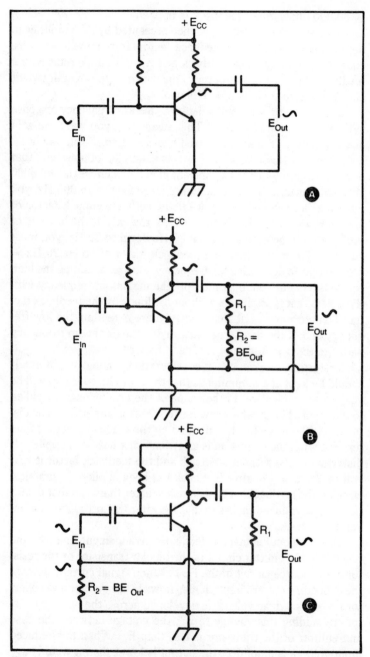

Fig. 10-1. A. Circuit without feedback. B. Circuit with negative or inverse feedback. C. Circuit in B, redrawn.

277

= 30,000 ohms when feedback is applied.

Similarly, the output impedance presented by the amplifier to the load is *reduced* by the feedback factor. If its impedance were equal to 6 ohms without feedback and if we used a circuit with a feedback factor of 3, the output impedance of the overall circuit has dropped to 6 ohms/3 = 2 ohms.

In audio work, you want a flat high- and low-frequency response over the entire audible band. This means that you want the gain of the circuit to remain at a fixed constant figure from about 20 Hz up to 15 kHz or 20 kHz. For this example, let's assume that our original amplifier has a constant gain between 80 Hz and 6000 Hz. If you want to extend the upper frequency to 15,000 Hz, you must add a negative feedback circuit with a feedback factor of 15,000 Hz/6000 Hz = 2.5. Similarly, if the gain at the low end of the band is to be extended from 80 Hz down to 20 Hz, you must add a feedback circuit with a feedback factor of 80 Hz/20 Hz = 4. Note that to determine the frequency you are extending the high end of the band, you must multiply the maximum frequency with fixed gain that is attainable without feedback by the feedback factor. To determine what frequency you are extending the low frequency end of the band, you must divide the minimum frequency with the fixed gain by the feedback factor.

Distortion is an important consideration in audio amplifiers. It should be kept at a minimum so that the reproduced signal will be very close or identical to the shape of the original signal applied to the input. This is where inverse feedback is important. The distortion without feedback is divided by the feedback factor to determine what the distortion is after feedback has been applied. If distortion is 3% without feedback and the feedback factor is 3, it will be 3%/3 = 1% after feedback has been applied. In practical circuits, the feedback factor is much greater than 3 so that distortion can be reduced to less than 0.1% after the introduction of inverse feedback.

A very popular inverse feedback arrangement involves the resistor in the emitter circuit of the bipolar transistor or the resistor in the source circuit of the FET. When signal current flows in the collector-emitter circuit, it also flows in the emitter and collector resistors. An ac voltage is developed across the emitter resistor. By adding this voltage to E_{In}, the voltage between the base and emitter of the transistor is less than E_{In}. When this reduced input voltage is amplified, the output is lower than it would have been had no ac voltage been developed across the emitter resistor.

This is illustrated in Fig. 10-2. Here an FET is being used rather than a bipolar junction transistor, but the same theory applies.

Note the relative polarities of E_{In} and the voltage across R_S. When added, less voltage remains to be applied between the gate and source than if R_S did not exist. As pointed out in a previous example, the effect of R_S on gain of the overall circuit can be cancelled if it is bypassed by a sufficiently large capacitor. When this is done, there is essentially a short circuit between the source and ground as far as the signal is concerned. But if this resistor or the equivalent resistor in the bipolar junction transistor is left in the circuit without the shunting capacitor, you have all the advantages that inverse feedback can provide.

In some circuits, the output voltage from a multistage amplifier is fed back through a resistor to the source of an FET. This feedback voltage is developed across R_S. If it is the proper phase, it can be used as the negative feedback required to improve the performance of the circuit. It was this type of feedback that was shown in Fig. 8-19B to set the frequency characteristics of the phonograph amplifier.

POSITIVE FEEDBACK

If there is positive feedback so that the signal fed back from the output to the input of the circuit is in phase with the input voltage, the gain of the overall circuit is greater than it is without this feedback. A circuit with positive feedback is shown in Fig. 10-3. This is not unlike the negative feedback circuit in Fig. 10-1. Just

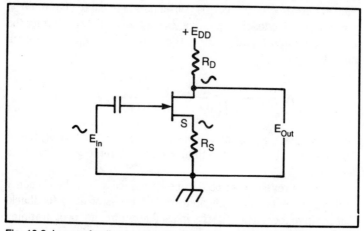

Fig. 10-2. Inverse feedback is developed because R_S is in the source circuit.

one more transistor amplifier stage, Q_2, is added to the circuit before feedback is applied.

As is always the case, the input signal between the base and ground of a transistor is 180° out of phase with the output signal between the collector and ground. This was described for the inverse feedback circuit and is equally true for the transistors in the circuit in Fig. 10-3. Signals at the input and output of Q_1 are 180° out of phase with each other. Signal at the output of Q_1 in Fig. 10-3 is applied between the base and ground of Q_2. When its amplified version appears at the collector of Q_2, its phase has been shifted 180° by this transistor. Thus, the phase of E_{In} has been shifted 180° to the collector of Q_1 and another 180° to the collector of Q_2 for a total phase shift of 360° from the input to the output of the circuit. As you know from the discussion of a radius rotating in a circle, a 360° shift puts the two signals—the input and the output—in phase with each other. The output voltage from Q_2 is divided between R_1 and R_2. The voltage across R_2 is added to E_{In} for a larger input voltage between the base and ground of Q_1 than would exist if only E_{In} was in the circuit. Because the transistor circuit has the same gain regardless of the size of the voltage at its input, the gain of the overall circuit is greater when positive feedback has been added.

For example, assume as before that E_{In} is 0.1 volt and the gain of the circuit without feedback, A_v, is 60. Without feedback, the output due to the gain is (0.1 volt) 60 = 6 volts.

Now assume R_2 in Fig. 10-3 is 1000 ohms and R_1 is 99,000 ohms. B for Equation 10-1 is equal to 1000 ohms/(1000 ohms + 99,000 ohms) = 1/100. BA_v is consequently equal to (1/100) 60 = 0.6. Because feedback is positive, BA_v is positive. Substituting the information into Equation 10-1, the gain of the positive feedback circuit is

$$A_f = \frac{60}{1 - (+0.6)} = \frac{60}{0.4} = 150$$

so that with E_{In} equal to 0.1 volt, the output is 0.1 volt × 150 = 15 volts. This is 15/6 = 2.5 times greater than the gain is without feedback.

The improvements noted in performance due to the presence of inverse feedback are negated when there is positive feedback. But positive feedback is useful in oscillator circuits, bootstrapping circuits, and so on.

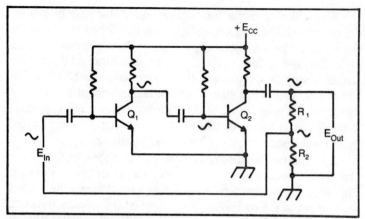

Fig. 10-3. Circuit with positive feedback.

Oscillators

Let's see what happens when one resistor in the positive feed-back circuit in Fig. 10-3 is changed. Instead of letting R_1 equal 99,000 ohms, let's make it equal to 59,000 ohms. Now B = 1000 ohms/(1000 ohms + 59,000 ohms) = 1/60. As a result, BA_v is equal to (1/60) (60) = 1. Substituting this into Equation 10-1, you will find that the gain of the circuit with positive feedback is equal to

$$A_f = \frac{60}{1 - (+1)} = \frac{60}{0} = \infty$$

because any number divided by zero is equal to infinity. With an infinite gain, there would be an output from the circuit even if the input voltage were extremely close to E_{In} = 0 volt. This is be-cause 0 volt (or any voltage) multiplied by an infinite gain gives you the maximum output the circuit can deliver.

Now suppose you have a feedback amplifier with infinite gain. At first, no voltage is applied to power it. The instant voltage is applied, there is a slight pulse at the input because current starts to flow. That slight pulse is amplified and results in an amplified voltage at the output. When a portion of that amplified voltage is fed back to the input, that portion of voltage is magnified by the transistor circuit to provide the peak output voltage that the cir-cuit is capable of delivering. This feedback and amplification con-tinues indefinitely. There is an output even though no external signal is placed at the input. When this happens the transistor is said to be in a state of oscillation.

An oscillating circuit provides an output that may be sinusoidal or any other shape, depending upon the circuit involved. But there are cyclic variations regardless of the waveshape of each cycle. Because of these cyclic variations, oscillators produce waveshapes at specific frequencies. If components are such that the circuit in Fig. 10-3 is in a state of oscillation, the frequency is determined by the magnitudes of the various resistors and capacitors in the circuit as well as by the resistances and capacitances formed by the semiconductor slabs inside the transistor.

A circuit can be designed to oscillate at a specific frequency. One simple oscillator using resistors and capacitors, is shown in Fig. 10-4. Ac voltage across R_C differs by 180° from the voltage at the input of the transistor. If you applied the voltage across R_C directly to the base of a transistor after it has passed through a voltage divider network, you would have the circuit in Fig. 10-1C. This is an inverse feedback circuit and it will not oscillate. To make this circuit oscillate, the signal fed back to the input must be in phase with what it is at the input. This occurs, of course, if the phase shift from input to output of the circuit is a multiple of 360°.

Here there is already a phase shift of 180° between the input and collector of the transistor. RC networks also provide phase shifts, as you saw in Chapter 5. At one frequency, the phase shift is 60°. This can be seen through the use of Fig. 5-4F, after X has been subtracted from X_L. When this has been done, you end up with a diagram similar to the one in Fig. 5-15B. This can be redrawn as in Fig. 10-5 for one RC network. In Fig. 10-5A, you see the tangent of the angle θ is equal to $(1/6.28\ f_c)/R = 1/6.28\ f_o RC$. This is the angle by which the phase was shifted.

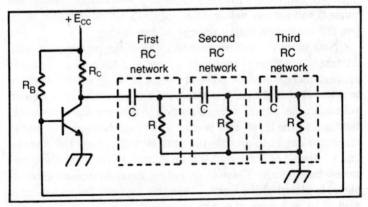

Fig. 10-4. Phase-shift RC oscillator.

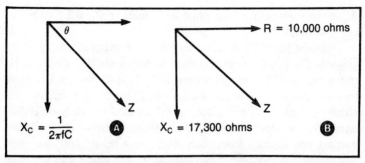

Fig. 10-5. Phasor diagrams for series RC network. A. Diagram showing phase relationships for RC network. B. Numbers are used in the diagram for the problem when f_o = 1 kHz and C = 9.2 nF.

Continuing with this analysis, when the phase shift is 60°, we find that tan 60° is equal to 1.73. Then $1/6.28f_o RC$ for the phasor diagram must also be equal to 1.73 when θ = 60°. For example, assume you want the 60° phase shift to be at 1000 Hz. If you made R equal to 10,000 ohms, $1/6.28f_o C$ must be equal to R tan θ = 10,000 (1.73) = 17,300 ohms. Because f_o = 1000 Hz, then

$$\frac{1}{6.28 \ (1000 \ Hz) \ C} = 17,300 \ ohms$$

and C = 9.2 nF. These figures are shown in Fig. 10-5B.

If you put three of these networks in series as in Fig. 10-4, you should get a total phase shift of 3 × 60° = 180° at one frequency. When this is added to the 180° phase shift provided by the transistor, there is a total phase shift of 360° from the input to the output. After the phase has been shifted 360° by the elements in this circuit, the output from the overall circuit is fed back to the input. The circuit oscillates at the one frequency where the phase shift of the overall circuit is 360°.

For this circuit to oscillate at any frequency, BA_V in Equation 10-1 must be equal to +1 or more. This may be a problem because each RC network is a voltage divider. For example, consider the first RC network in the circuit. The output voltage from across R_C is divided between this R and C. A portion of the total voltage from R_C is across R. This is divided further by the second and third RC networks. From there, it is fed back to the base of the transistor. B in the equation is the ratio of the voltage fed back to the input to the voltage across R_C. The ratio multiplied by the voltage gain

A_v of the transistor circuit must be at least equal to 1 for the circuit to oscillate.

Assume that BA_v is somewhat above 1 in our circuit, so it oscillates. You can next assume that the circuit oscillates at 1000 Hz if you put three RC networks, each identical to the one analyzed, into the circuit. From our analysis of the oscillator circuit, this should happen with capacitors of 9.2 nF and resistors of 10,000 ohms. But this does not happen. The three RC networks in the circuit are not isolated from each other. As a result, the three networks interact. Because of this interaction, the phase shift provided by the combination of the three networks differs from the 180° shift predicted that was at frequency f_o. If all resistors and capacitors are the same as the ones used in the analysis of the circuit in Fig. 10-5B, the actual frequency provided by the oscillator would differ from the predicted 1000 Hz. An approximation to the actual frequency of oscillation can be determined through use of Equation 10-2.

$$f_o = \frac{1}{18 \text{ RC}} \qquad \text{Eq. 10-2}$$

Substituting in the numbers, $f_o = 1/18 \ (10^4) \ (9.2 \times 10^{-9}) = 604$ Hz.

Three RC networks are shown in Fig. 10-4; each one shifts the phase by about 60° for a total shift of 180°. Four RC networks are used in some circuits, which oscillate so that the total phase shift by the RC networks is still 180°. Thus, each RC network shifts the phase by about 180°/4 = 45°. The same phase-shift discrepancies exist here as in the three RC network circuit because of the interaction between the networks. Some networks shift the phase by more than 45° and others by less than 45°. The total is still a 180° shift by all four RC networks at the frequency of oscillation. That frequency is approximately equal to

$$f_o = \frac{1}{9RC} \qquad \text{Eq. 10-3}$$

For example, using a four-RC-network oscillator, assume R = 15,000 ohms and C is equal to 10^{-8}F (or 0.01 μF). Now $f_o = 1/9$ (15,000) $(10^{-8}) = 741$ Hz.

On some oscillators, not all R's and C's are identical. If you should encounter one of these oscillators, there is no way you can use either Equation 10-2 or 10-3 to determine the generated frequency. It can only be determined through use of test instruments such as an oscilloscope or a frequency counter.

Another type of oscillator using an RC network is shown in Fig. 10-6. Here the phase of the input voltage is shifted 180° by Q_1 and then another 180° by Q_2 so that the voltage across R_{D2} is 360° out of phase or 0° in phase with the input at Q_1. If the drain of Q_2 were connected though an isolating capacitor directly to the gate of Q_1, this circuit could oscillate, but the frequency can not be determined by the obvious components in the circuit.

To set the frequency of oscillation, a *Wien bridge* circuit is added between the input and output, which consists of two identical resistors and two identical capacitors. Once the magnitudes of these components are known, the frequency of oscillation can be determined using Equation 10-4.

$$f_o = \frac{1}{6.28RC} \qquad \textbf{Eq. 10-4}$$

The series RC network in the Wien bridge circuit forms a voltage divider with the parallel RC network. At the oscillating frequency, only one-third of the voltage across the combination of the two networks remains across the parallel network. Because of this, the gain of the overall circuit must be at least equal to 3 to compensate for this loss. If gain is less than 3, the circuit will not oscillate.

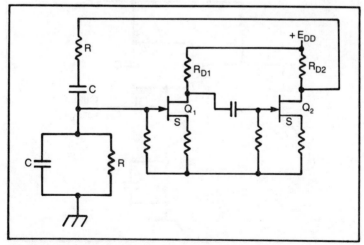

Fig. 10-6. Wien bridge oscillator.

Wien bridge oscillators are used in audio signal generators supplied by different instrument manufacturers. These generators can usually provide a wide range of frequencies. Fine variation of frequency is accomplished by adjusting the two identical C's through use of a two-gang variable capacitor. The range or coarse change in frequency is made by using a switch to select different resistors for R.

Most rf oscillators in use today use LC networks for selecting the frequency. A parallel inductor-capacitor circuit resonates at one frequency so the impedance it presents is high only at that resonant frequency. One such circuit is shown in Fig. 10-7. Here, the LC resonant circuit presents its highest impedance in the collector circuit at the resonant frequency. The signal developed in this tank circuit is induced into the coil winding marked L_1. The induced voltage across that coil appears between the base and ground of the transistor and is amplified by the transistor circuit. The magnified output appears across the resonant circuit and is once again induced into L to keep repeating the amplifying process. After several cycles, a fixed output equal to the resonant frequency of the tank circuit, $1/6.28 \, (LC)^{1/2}$, is across the tank. Voltage at the oscillating frequency can be taken from between the collector terminal of the transistor and ground and fed to any other circuit.

You might ask how you can be sure that the voltage at the input and the output to the transistor are in phase. Considering the tank

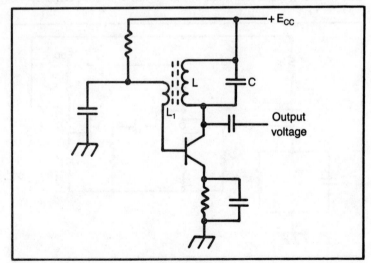

Fig. 10-7. Oscillator using resonating LC circuit.

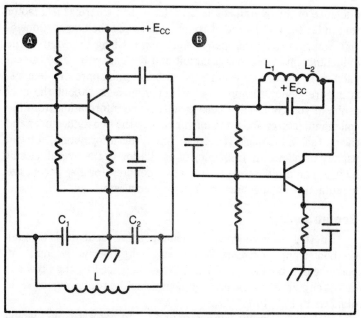

Fig. 10-8. Resonant circuits used in oscillators. A. Colpitts circuit. B. Hartley circuit.

circuit and the signal induced into L_1, the two voltages can be either in phase or 180° out of phase with each other. If they are not in phase, the circuit will not break into oscillation. All that need be done is to reverse the connections to L_1. This alters the relative phases of the windings by 180°.

There are numerous variations on this LC oscillator circuit. The most popular are the Hartley and Colpitts circuits shown in Fig. 10-8. In the Colpitts circuit, the two capacitors that are across the inductor are connected in series. To determine their equivalent capacitance, you can use Equation 3-2, substituting the C's for the R's in the equation. Then the resonant frequency can be found through use of Equation 5-12. For example, if C_1 = 100 pF and C_2 = 50 pF, the equivalent capacitance of C_1 in series with C_2 is $(C_1)(C_2)/(C_1 + C_2)$ = (100 pF)(50 pF)/(100 pF + 50 pF) = 33.3 pF. If L is 1 mH, then the resonant frequency of the LC circuit and the frequency of oscillation is $1/6.28\sqrt{LC}$ = 1/6.28 [$(10^{-3}$H)$(33.3 \times 10^{-12}$F)]$^{1/2}$ = 873 kHz.

In the Hartley circuit in Fig. 10-8B, the inductance is split into two sections. The point where it is split gets connected to the supply voltage E_{CC}. This terminal is at ac ground. The equivalent inductance of the overall coil is $L_1 + L_2 + 2M$, where M is a factor that indicates the degree of coupling between the two sections of coil, referred to as the *mutual inductance*. Using this sum as the inductance in the resonant circuit and the capacitor C as the capacitance, you can determine the resonant and approximate oscillating frequency through use of Equation 5-12. About the only problem you could have here is setting the information from the coil manufacturer about the magnitudes of the two inductances and the mutual inductance. If you cannot get this information, you can usually measure the frequency of oscillation of the overall circuit by using an oscilloscope or frequency counter—or else accept the manufacturer's specification as being accurate.

Bootstrapping

Another positive way of utilizing positive feedback is through the bootstrapping circuit. When designed properly, the input impedance of a bootstrapped amplifier is increased by the presence of this circuit. A typical arrangement using the bootstrapping technique is in Fig. 10-9.

If R in the circuit were zero and C did not exist, you would have the usual common-emitter amplifier. The impedance seen by E_{In} is R_B in parallel with R_x. These components are in parallel with βR_E.

In this circuit, C is made very large and feeds voltage back from the emitter to the base of the transistor, through R. Because the voltages at the emitter and base are in phase, the feedback through C is positive. Being very large, C behaves as if it is a short circuit for the signal voltage. If it were connected directly to the base, it would short the base to the emitter. Instead, it is connected to one end of R, and R isolates C from the base.

One end of R is connected to E_{In}. A voltage almost as large as E_{In} exists across R_E just as in any common-collector circuit. The voltage across R_B is applied through capacitor C to the second end of R. Now the same voltage is at both ends of the resistor. Because there is no (or very little) difference of potential across R, no current (or an infinitesimally small amount) flows through it. As a result, R is equal to this slight voltage divided by essentially zero amperes of current. Using this information in Equation 2-2, you will find that the resistance of R appears to be almost infinite. Because this resistance is large, it has no significant shunting effect on the input of the circuit.

Because of the feedback through C, R_B, and R_x are in the

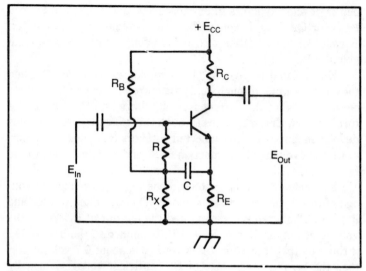

Fig. 10-9. Bootstrapping circuit.

emitter circuit. The overall impedance to the input circuit appears as a resistance equal to the beta of the transistor multiplied by the parallel combination of R_B, R_x, and R_E.

For example, if R_B = 50,000 ohms, R_x = 10,000 ohms and R_E = 500 ohms. Without positive feedback, E_{In} sees R_B in parallel with R_x or (50,000 ohms) (10,000 ohms)/(50,000 ohms + 10,000 ohms) = 8,333 ohms (using Equation 3-2), and this resistance is in parallel with βR_E. If β = 20, then βR_E = 10,000 ohms, so the total resistance seen by E_{In} is (8333 ohms) (10,000 ohms)/(8333 ohms + 10,000 ohms) = 4550 ohms.

Using positive feedback, the 8333 ohms is in parallel with the 500 ohms for an equivalent resistance of (8333 ohms) (500 ohms)/(8333 ohms + 500 ohms) = 471 ohms. Now E_{In} sees 471 ohms multiplied by the beta of the transistor, or 9420 ohms. This is more than twice the 4550 ohm impedance E_{In} sees without the bootstrapping circuit.

VOLTAGE REGULATORS

In earlier chapters I discussed voltage regulators using zener diodes, but no feedback circuits were involved. Modern regulated supplies do use negative feedback circuits to improve regulation. A basic circuit of this type is shown in Fig. 10-10.

Unregulated but filtered dc voltage from a full-wave supply circuit is developed across filter capacitor C and applied to the collector or Q_2. If Q_2 is conducting, a dc voltage is at its emitter. Because of the feedback circuit involving both Q_1 and Q_2, that voltage, V_{Reg}, is fixed.

Unregulated voltage, V_{Unreg}, is applied to the base of Q_2 and the collector of Q_1 through R_1. If the current through Q_1 is not excessive, some of the current from R_1 flows into the base of Q_2, turning it on. Current passes through this transistor. A voltage is developed across series-connected resistors R_3 and R_4. The portion of the voltage at the junction of R_3 and R_4 is fed back to the base of Q_1.

R_2 conducts current to zener diode D_3. The breakdown voltage of this device is across the diode. This is a fixed voltage and it appears between the ground and the emitter of Q_1. When the voltage at the junction of R_3 and R_4 is applied to the base of Q_1, it turns on that transistor if this voltage is about 0.7 volt greater than the breakdown voltage of D_3. Sufficient voltage is usually available at that junction to keep Q_1 turned on at all times.

When the output voltage stays at its ideal level, the voltage at the base of Q_1 remains fixed. A specific current flows through the transistor. Should some condition occur to cause V_{reg} to jump up a bit, more voltage is available at the junction of R_3 and R_4 than before, turning Q_1 on harder than when V_{reg} is at its ideal level. Should V_{reg} drop for some reason, less voltage than the ideal is available for the base of the transistor, so Q_1 conducts less current than it would under ideal conditions.

Although collector current for Q_1 as well as base current for Q_2 flows through resistor R_1. The total current through this resistor does not change. It is fixed regardless of the portion of the current that flows into the collector of Q_1. So if V_{reg} jumps up, causing more collector current than normal to flow in Q_1, less current is left for the base of Q_2. This reduction of current in the base means that less current flows in its emitter and in the load, R_L. If less current flows here, the voltage across the load is reduced and the output drops to the ideal V_{reg} volts.

Should the opposite situation exist and V_{reg} drop below the ideal level, less current flows in the collector of Q_1 and more current is available for the base and hence emitter of Q_2. This added current through the load forces V_{reg} back up to its ideal magnitude.

This type of regulator is quite stable despite line voltage and load variations. There are several improvements possible over the

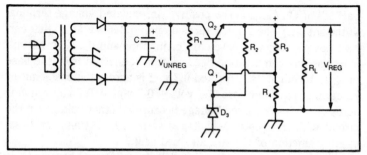

Fig. 10-10. Feedback regulator circuit.

circuit shown in the figure. The most important improvement is in Q_2. Ideally, its beta should be large. One transistor is usually insufficient to supply the required beta for the circuit. In many circuits, two transistors are used in a Darlington configuration to replace the individual Q_2 shown in the diagram. The beta of the combination is quite large for it is the product of the betas of the two devices.

A second improvement involves a *preregulator*. This is a circuit that provides a fixed current rather than voltage, and is used to replace R_1 in Fig. 10-10. The input circuit to the base of Q_2 and the collector of Q_1 now looks like the one in Fig. 10-11. Here current flows through the series circuit consisting of D_4, D_5, and R_5. As always, a fixed 0.7 volt is across each of the forward-biased diodes, so that there is 1.4 volt between the base of Q_3 and termi-

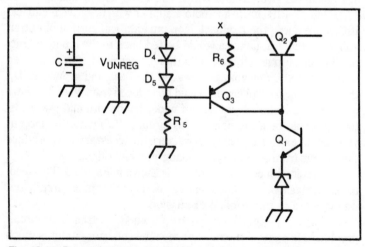

Fig. 10-11. Preregulator circuit added to regulate power supply.

nal "x" of Q_2. (This is the positive unregulated voltage terminal of the supply.) The 1.4 volt turns on pnp transistor Q_3 so that current flows through its base-emitter junction and R_6. Now 1.4 volt is across this combination because it is connected from the base of Q_3 to terminal "x." A fixed 0.7 volt is across the base-emitter junction of Q_3. Therefore, 1.4 volt − 0.7 volt = 0.7 volt remains across R_6. This is a fixed voltage because all other voltages in the circuit with this resistor—voltages across D_4, D_5, and the base-emitter junction of Q_1—are all fixed at 0.7 volt.

The total current fed to the collector of Q_1 and base of Q_2 is from the collector of Q_3. But this is just about equal to the current in R_6. Using Ohm's law, that current is fixed at 0.7 volt/R_6. Thus, the circuit involving Q_3 is a *constant-current* source.

This constant current is desirable in the power supply circuit for several reasons. For one, V_{Unreg} may have some ripple. If there is a constant current flowing into the base of Q_2, that ripple never reaches the emitter of that transistor nor the load at the output. In a similar manner, variations in the unregulated voltage will not appear as such at the output. This circuit therefore makes it easier for the feedback regulator portion of the circuit to do its job.

FEEDBACK AMPLIFIERS

The different methods of applying negative feedback to an amplifier were noted earlier in the introduction to this chapter. Feedback is used in many different circuits. In the discussion in Chapter 8, I discussed two circuits which can or do use feedback.

Figure 8-19B shows an audio amplifier with negative feedback. If C_1 were not in the feedback loop, an identical portion of output voltages at all frequencies would be fed back to the input. A fraction of this output voltage would then be across R_{E1}. When added to the applied input voltage between the base and ground of Q_1, the gain of the overall circuit would be reduced from what it is without feedback. The advantages of wider bandwidth and lower distortion would be a big plus in this circuit. With C_1 in the circuit, you add an extra frequency characteristic necessary to reproduce the output from a magnetic phonograph cartridge.

A circuit with positive feedback is shown in Fig. 8-25. This feedback involves C_1, a component in the bootstrapping circuit that was ignored in the original discussion.

A large voltage gain is required from Q_1. A transistor circuit provides a big gain if the resistance in the collector circuit is much larger than the resistance in the emitter circuit. Thus, Q_1 will provide a large gain if the resistance in the collector circuit of the transistor is large. C_1 feeds back signal from the output to the junction of R_{B2} and R_{B3}. These two signals are normally in phase, so C_1 performs as if it is in an ordinary bootstrapping circuit. R_{B2} appears much larger in the collector circuit of Q_1 now, than it would if C_1 did not exist.

C_1 also serves a second function but it has nothing to do with feedback. During the portion of the cycle when the current through Q_4 is large, very little voltage is across this transistor. Its emitter and collector voltages are just about identical and equal to $+B_{CC}$. At this time, the base and emitter of Q_2 are also at the same $+E_{CC}$ voltage, turning off this transistor. A distorted output is therefore present during that portion of the cycle. Here is where C_1 comes into play.

When the circuit is idling, C_1 is charged. This voltage is between the emitter of Q_2 and the base of Q_2, through R_{B2}. When the emitters of Q_2 and Q_4 are brought to E_{CC} volts, the base of Q_2 is still turned on due to the voltage across the previously charged capacitor C_1. Q_2 is turned on because this voltage across C_1 keeps its base positive with respect to its emitter. Now Q_2 conducts at the peaks of the signal. When it conducts, it also turns on Q_4. Distortion is thereby eliminated during this peak portion of the cycle.

Negative feedback is almost always present in this circuit, even through it is not shown in Fig. 8-25. A resistor is usually connected from the junction of C_L and R_L to the base of Q_1. Voltage is fed back through this resistor from the output of the amplifier to its input. Ac voltage across the load is out of phase with the input signal, so the feedback is negative. Voltage from across R_L is divided between the feedback resistor and the ac input resistance presented by the Q_1 circuit. Thus, the amount of feedback in a circuit is determined by the magnitude of the feedback resistor.

There are a number of not too obvious phase shifts in this type of circuit. The shift that occurs due to the presence of the $C_L = R_L$ circuit is pretty obvious because there is a resistor and a capacitor in the circuit. But there are capacitances between the semiconductor materials in each transistor. This capacitance is small, so it has an effect primarily on phase shifts at high frequencies. Similarly, a loudspeaker has an inductance. That inductance, coupled with the resistance in the balance of the circuit, causes a phase shift. This shift is usually significant at somewhat above the audio frequency range. Because of these phase shifts, the voltage across R_L is in phase with the voltage at the input of Q_1 at some frequencies. Thus, the presence of the feedback resistor can cause some amounts of positive feedback to exist and the circuit may oscillate.

In order to minimize the chances of oscillation at these higher frequencies, a small capacitor is placed across the feedback resistor. It shifts the phase of the high-frequency signals when they are fed back so that feedback is negative even at these frequencies. The stability of the overall circuit depends upon the capacitance of the component used as well as on the magnitude of the feedback resistor. These values should never be changed on any amplifier, but the components should be checked if the amplifier oscillates or distorts the signal.

OPERATIONAL AMPLIFIERS

Integrated circuits housing *operational amplifiers* or op amps, have become extremely popular. These devices serve as amplifiers in feedback circuits. You don't need to know the details about the circuitry inside these ICs. This is because each one is a closed "chip" that does a particular job. You will, however, see what this device can do.

The schematic representation of an op amp is shown in Fig. 10-12. Usually equal positive and negative supply voltages are applied to the device. Although these are shown here as $+E_{CC}$ and $-E_{CC}$, these leads are usually omitted from the schematic.

In this diagram, there are two inputs. The *inverting* input is labelled $-$ and the *noninverting* input is labelled $+$. Any signal fed to the $-$ input is magnified and inverted 180° before it appears at the output. Should the positive half of an ac cycle be fed to the $-$ input with respect to the $+$ input, a negative half-cycle is between the output and ground. As for the $+$ input, the output is in

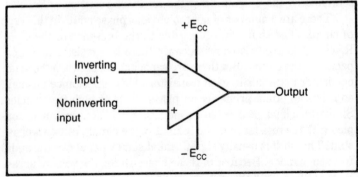

Fig. 10-12. Schematic representation of an op amp.

phase with any signal applied to that point. If a resistor is connected between the output and the + input, the circuit may oscillate because of the positive feedback. Similarly, gain is reduced due to negative feedback if a resistor is connected between the output and the inverting input.

Voltage Comparator

The maximum positive peak output voltage from an op amp is $+E_{CC}$ volts, while the maximum negative peak is $-E_{CC}$ volts. Obviously, the output peaks from an op amp or any other amplifier can never exceed the supply voltage.

Op amps also have extremely large gain when no negative feedback is applied to the circuit. Thus, a tiny positive voltage at the – input would push the output to $-E_{CC}$ volts and a tiny negative voltage at this input would push the output to $+E_{CC}$ volts. Similarly, a tiny positive voltage at the + input results in $+E_{CC}$ volts at the output, while a tiny negative voltage at this terminal causes $-E_{CC}$ volts to be at the output. The polarity of all input voltages is with respect to the common or fixed voltage at the alternate input terminal.

Using this information, you can see how the circuit in Fig. 10-12 can perform as a voltage comparator.

Assume that the + input is connected to ground. If the – input is also connected to ground, there is zero voltage at the output of an ideal op amp. Should a slightly positive voltage be applied to the – input, the voltage at the output drops to $-E_{CC}$. Similarly, a slight negative voltage at this input would bring the output to $+E_{CC}$ volts. The polarity of the output voltage is due to the 180° phase shift from the – input to the output. The magnitude is due to the enormous amplification capabilities of this circuit. But note the voltage comparator application. The output voltage is an indication that the voltage at the – terminal differs from the 0 volt or round voltage applied to the + terminal.

Now suppose that the + terminal is fixed at +3 volts. This can be done by putting a battery between the + terminal and ground or by applying a rectified and filtered voltage to this terminal from a power supply. Now +3 volts at the – terminal would put identical voltages at the two input terminals. The ideal output voltage is zero. A slight drop below +3 volts at the – terminal, let us say to +2.99 volts, makes the minus terminal more negative than the + terminal. Because of the phase relationship, the output is now +Ecc volts. If the – terminal were at +3.01 volts or slightly more positive than the + terminal, the output would be $-E_{CC}$ volts.

This is the basis of the voltage comparator, which simply compares the voltages at the two input terminals. If there is a voltage at the output, you know that the two input voltages differ from each other. The polarity of the output voltage tells you if the voltage at the − terminal is greater than or less than the voltage at the + terminal. And all this is accomplished by the circuit without *external* feedback. The only feedback is inside the chip.

Instead of fixing a voltage at the + terminal and comparing the voltage at the − terminal with this voltage, voltage at the − terminal can be fixed and changes in voltage at the + terminal can be compared with it. If the − terminal were fixed at +3 volts, any voltage greater than +3 volts at the + terminal would result in an output of $+E_{CC}$ volts because the + terminal is more positive than the − terminal. Using this same logic, if the voltage at the + terminal dropped even slightly below +3 volts, it would be negative with respect to the voltage at the − terminal. Now the output would be $-E_{CC}$ volts.

Amplifiers

The enormous gain of the op amp is reduced when it is used as a voltage amplifier. In this type of circuit, negative feedback is added around the IC. This is shown in Fig. 10-13. In both circuits, output voltage is fed back through R_f to the inverting input. In Fig. 10-13A, the input signal is applied to the − input. Here, the output is 180° out of phase with the input. In the circuit in B, E_{In} is applied to the + input of the op amp. Output and input voltages are in phase with each other. The gains of the two circuits illustrated here are different. The voltage gain of the inverting circuit is

$$A_V = \frac{R_f}{R} \qquad \text{Eq. 10-3}$$

while the gain of the noninverting circuit is the same as shown in Equation 10-3, with the addition of a 1.

For example, if R_f = 40,000 ohms and R = 10,000 ohms, the voltage gain of the circuit in Fig. 10-13A is 40,000 ohms/10,000 ohms = 4. So if E_{In} were 0.3 volts rms, the output is 4 × 0.3 = 1.2 volts rms. For the circuit in Fig. 10-13B, the voltage gain is the same as in the circuit in B with the addition of a 1. Here the voltage gain is 4 + 1 = 5. With E_{In} equal to 0.3 volt rms, the out-

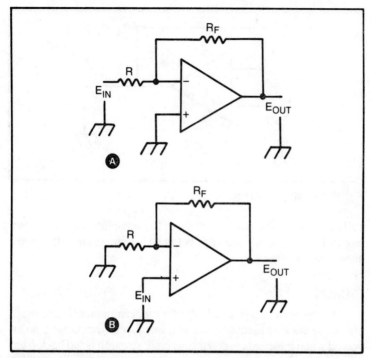

Fig. 10-13. Op-amp voltage amplifiers. A. Inverting amplifier. B. Noninverting amplifier.

put voltage from this circuit is $5 \times 0.3 = 1.5$ volts rms.

It is interesting to note that when feedback is used in the circuit, voltages at the two input terminals of the op amp are just about identical. So if the $+$ input is at ground, the $-$ input terminal is close the the ground potential and is said to be at a *virtual ground*. But this does not affect the actual voltage gain of the overall circuit.

Remember the emitter-follower using a bipolar transistor and the source-follower or common-source circuit using an FET? A voltage follower with characteristics similar to that of a follower using a transistor can be made instead with an op amp and is shown in Fig. 10-14. Here is a complete feedback circuit from the output to the inverting input. The output voltage is equal to the input voltage and is in phase with it.

Other Op-Amp Circuits

Many different arrangements are possible with the op amp. There are adders, subtractors, feedback filter circuits, oscillators,

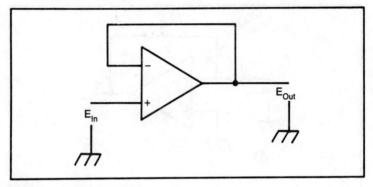

Fig. 10-14. Voltage follower.

and so on. Typical circuits for each of these applications are shown in Fig. 10-15. The relationships between the inputs and outputs are shown in the figure.

RADIOS

Oscillators are typically built into modern radios. These oscillators use positive feedback circuits. Feedback circuits are also involved when information about the signal strength is fed back from one section of the electronics in the radio to another section. This information is obtained at the audio portion of the circuit and is fed back to amplifiers nearer the input stage of the radio. The data fed back is in such form as to keep modifying the overall gain of the radio as the magnitude of the received radio signal changes. In this way, the audio output from the radio remains relatively constant despite changes in the strength of the received signal. There are also negative feedback circuits in the audio amplifier portion of the radio.

Amplitude Modulation

The frequency of radio signals varies from a low 20,000 Hz up to many billions of Hertz or gigahertz. Here, we are only concerned with the broadcast band which ranges from 535 kHz to 1605 kHz. A different radio frequency is transmitted in this range by each station. Due to the characteristics of the transmitted signal, an individual frequency is not sufficient for each radio station. Each station is therefore allotted a range of frequencies for its use. This range can be as low as plus and minus 5000 Hz around its assigned center frequency, or may be as high as plus and minus 10,000 Hz.

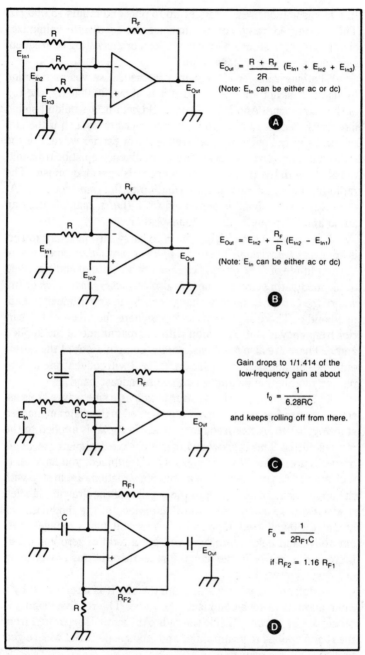

$$E_{Out} = \frac{R + R_F}{2R} (E_{In1} + E_{In2} + E_{In3})$$

(Note: E_{In} can be either ac or dc)

A

$$E_{Out} = E_{In2} + \frac{R_F}{R} (E_{In2} - E_{In1})$$

(Note: E_{in} can be either ac or dc)

B

Gain drops to 1/1.414 of the low-frequency gain at about

$$f_0 = \frac{1}{6.28RC}$$

and keeps rolling off from there.

C

$$F_0 = \frac{1}{2R_{F1}C}$$

if $R_{F2} = 1.16 R_{F1}$

D

Fig. 10-15. Circuits using op amps. A. Adder. B. Subtractor. C. Low-pass filter. D. Oscillator.

In a few isolated cases, it ranges up to plus and minus 15,000 Hz. This permitted range of frequencies or *bandwidth* minimizes changes of interference between stations broadcasting on frequencies that are close to each other.

Whatever radio or rf frequency is used, it serves no function by itself. It is not audible because it is about 20,000 Hz, the top of the audio band. And even if you could hear it, it wouldn't sound like much because there would be no music or speech if the station transmitted only the radio frequency or carrier wave. For the radio station to perform its function, intelligence must be transmitted along with the rf. This "intelligence" is speech or music. The radio frequency carries this information to the radio receiver. At the receiver the signal is separated from the rf, magnified by an audio amplifier, and fed to a loudspeaker.

The process of modifying the rf to carry the speech or music, is known as *modulation.* There is *amplitude modulation* or am where the amplitude of the carrier is changed by the size of audio applied to the modulating circuit. There is also *frequency modulation* or fm, where the carrier frequency changes with the amplitude of the applied audio. There is *phase modulation*where the phase of the carrier frequency is shifted in step with the magnitude of the applied audio. There are also *pulse modulation,* combinations of the different types of modulation, and so on. Here, I will concentrate on amplitude modulation as applied to the broadcast band.

An unmodulated rf transmitter broadcasts sine waves at its assigned carrier frequency. This is represented in the diagram marked rf in Fig. 10-16. A low frequency or audio signal is applied to the rf transmitter. This is shown by the sine wave designated as *audio* in the figure. When the two signals are combined, you have an rf with amplitude variations. The amplitude variations are in step with the applied audio, as shown in the section of the drawing labeled *rf modulated by audio.* Note the fluctuations of the modulated rf around the 0-volt level. If you add the plus and minus voltages present above and below zero, you will get a total of zero volts. But there is rms power in the signal just as there is rms power in an ordinary sine wave.

Modulated signal flows through the air. It is intercepted by a radio antenna or by a coil inside the radio. The received signal is processed by circuitry inside the radio, the audio is separated from the rf and applied to an amplifier and loudspeaker. But we are getting ahead of ourselves!

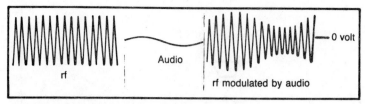

Fig. 10-16. Modulating rf with a sine wave.

Superheterodyne Radio

Modern radios utilize the *heterodyne* principle in their receive circuitry. Musically inclined readers have witnessed heterodyning when two different, very loud musical notes were played at the same time. You not only heard the two individual notes, but also two additional notes. The frequencies of the additional notes were equal to the sum and the difference of the two original notes. Superheterodyne radios utilize this heterodyne phenomenon in their circuits, but here the two rf signals that are combined produce frequencies other than audio tones.

A block diagram of a radio is shown in Fig. 10-17. Modulated rf is intercepted by the antenna, amplified in the *rf amp* and fed to a *mixer*. A local oscillator at the mixer is tuned to a frequency that is 455 kHz above that of the received rf. Output from the oscillator is fed to the mixer along with the amplified rf. The two signals are combined in the mixer. Just as in the case when the two audio notes were combined, four signals are at the output of the mixer. These are the received rf, the oscillator frequency, and the sum of the two frequencies and the difference of the two frequencies. For example, assume that a radio station broadcasting on 1000 kHz is being received. Because the local oscillator would be set to supply a frequency of 455 kHz above the frequency transmitted by the radio station, it oscillates at 1455 kHz. When the 1000 kHz is mixed ith the 1455 kHz, the heterodyned outputs from the circuit are at 1000 kHz, 1455 kHz, 1000 kHz + 1455 kHz = 2455 kHz and 1455 kHz − 1000 kHz = 455 kHz. Because the oscillator frequency is always 455 kHz above the frequency of the received rf, there is always a difference frequency of 455 kHz available. Because of this, all other frequencies at the output of the mixer are discarded. Only the 455 kHz is passed by the i-f or intermediate *frequency* amplifier to the balance of the i-f circuit. Because the rf used when forming this difference frequency is modulated by audio, the 455 kHz is modulated by the same audio.

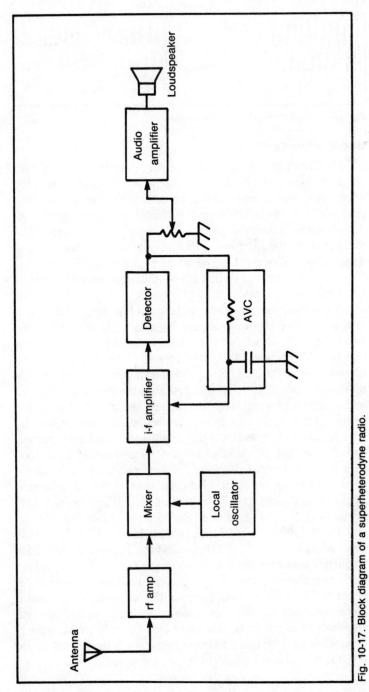

Fig. 10-17. Block diagram of a superheterodyne radio.

The modulated i-f signal is detected. Here the audio is separated from the rf, amplified in an audio amplifier, and applied to a loudspeaker.

The output from the detector is actually half of the modulated signal shown in Fig. 10-16, either the upper half of the signal above the 0-volt line or the lower half that is below the 0-volt line. When only one-half of the waveform is present as it is here, then the total voltage of the modulated signal is no longer equal to zero. The voltage of the upper or lower half of the modulated rf has an amplitude that varies in step with the audio that modulated it. If this signal were analyzed mathematically, you would find that there are actually rf and audio components present in the waveshape. To get only audio from the overall signal for application to the power amplifier and loudspeaker, the rf from the output of the detector is filtered and only audio remains. Audio is passed through a variable resistor or volume control and then applied to the input of the audio amplifier.

The magnitude of the rf output from the detector depends upon the size of the signal applied to the input of the radio circuit. To compensate for rf signal level variations, a circuit is provided which first passes the rf in the detected signal through a resistor and then applied to a filter capacitor. Because this signal is either above or below the 0-volt line, it has only one polarity and is an effectively pulsating dc. The pulses are eliminated through use of the filter capacitor in what is known as *avc circuit*. The magnitude of the filtered dc voltage is proportional to the strength of the received signal. The dc voltage is applied to the i-f amplifier to again bias the transistors in this circuit. This change of bias affects the gain. If the applied filtered voltage from the detector is relatively low, the overall gain of the rf amplifier is higher than it is when the filtered voltage is high. Because the gain is changed with changes in the rf output voltage from the detector, the output from the audio circuit remains at a fixed level regardless of the magnitude of the input rf signal. This *automatic volume control* (or *automatic gain control*) circuit is thus a type of feedback arrangement.

Circuit Description

No rf amplifier stage is included in the actual circuit of the radio shown in Fig. 10-18. Rf is fed directly from the secondary winding of the antenna coil to mixer Q_2. Q_1 is also in the local oscillator circuit. L_S, L_P, C_P, and Q_1 make up the oscillator. Signal developed across a portion of the oscillator tank circuit, C_P-L_P is fed to the

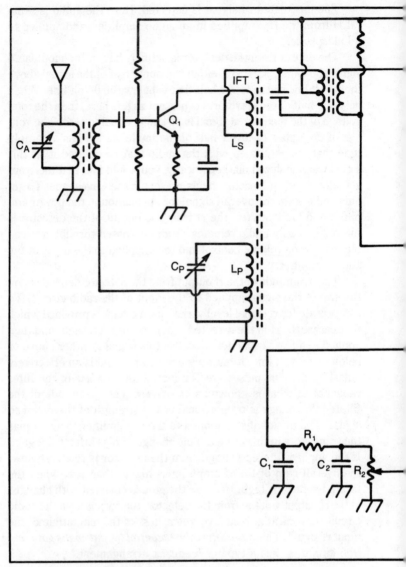

Fig. 10-18. Superheterodyne radio circuit.

secondary winding of the antenna coil. The total signal is amplified by Q_1 and applied through L_S to the primary of intermediate frequency transformer IFT-1 and is tuned to 455 kHz.

The local oscillator is in a feedback circuit extending from the input of Q_1 to its output. Signal from the output is fed back to the

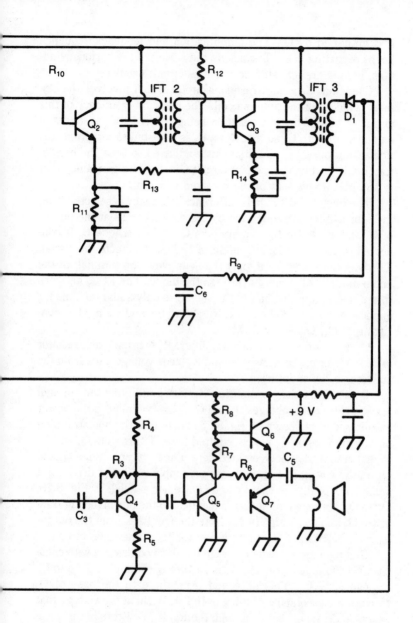

input. The local oscillator voltage at the input of Q_1 is amplified by this transistor. The amplified output is developed across L_S and is led from there into L_P in the resonant tank circuit. This circuit is used to set the oscillator frequency. A portion of the tank circuit voltage is fed back to the secondary of the antenna coil. From here it is amplified by Q_1. The amplification and feedback sequence keeps repeating itself. Because of the wiring of the various windings, phases are adjusted so that this circuit oscillates.

Capacitor C_A is in a resonant circuit which involves the primary winding of the antenna coil. The resonant frequency of this circuit is that of the radio station being received. C_A is mounted on one shaft with C_P so that when C_A is adjusted to receive a particular station, C_P is simultaneously adjusted to tune the oscillator circuit (involving resonant circuit C_P - L_P) to a frequency that is 455 kHz above that of the received rf signal.

The modulated 455 kHz difference frequency is selected from the four available frequencies by the various IFTs at the output of Q_1 and amplified by i-f amplifiers Q_2 and Q_3. Diode D_1 is connected to the secondary winding of IFT-3. It detects the signal. The way it's shown here, only the negative or bottom half of the audio-modulated rf passes through the diode. The R_1-C_2 network filters the rf from this half of the signal, so only audio remains for application to the level control, R_2, and to the audio amplifier consisting of Q_4, Q_5, Q_6, and Q_7.

Q_4 is an ordinary voltage amplifier. R_5 is a feedback resistor in the emitter of the transistor, and R_3 feeds voltage back from the collector of Q_4 to its base.

Q_5 is also a voltage amplifier. Besides amplifying the voltage, it also drives output stages Q_6 and Q_7. This is similar to the power amplifier circuit shown in Fig. 8-25 except that the bootstrapping circuit there using C_1 is not required here. Transistors Q_4 and Q_5 in that circuit have been omitted here. For a portable radio, Q_2 and Q_3, in the circuit in Fig. 8-25, would supply sufficient drive for a small loudspeaker so that Q_4 and Q_5 are not needed. In Fig. 8-25, Q_2 and Q_3 are in a circuit that is identical to the one involving transistor Q_6 and Q_7 in Fig. 10-18. Note the feedback from the output load to the base of Q_5 through resistor R_6 in the radio circuit.

Getting back to the detector circuit, the modulated rf with negative polarity is fed from the detector through R_9 and applied to filter capacitor C_6. This dc is fed from there to the base of Q_2 through the secondary winding of IFT-1. Without the voltage that is across C_6, Q_2 is biased through the R_{10}-R_{11} resistors in the base-emitter circuit. This bias sets the maximum gain level of the stage, the gain is modified when the voltage developed across C_6 is applied to the bias circuit. The magnitude of this voltage affects the voltage gain of Q_2. Because this avc voltage varies with the strength of the received signal, it varies the gain of Q_2 in step with the gain of the received rf. This is an attempt to keep the audio output at a fixed level regardless of the strength of the received rf.

The bias voltage for Q_3 is set by dc voltage at its base-emitter circuit due to the presence of resistors R_{11}, R_{12}, and R_{13}. But voltage across R_{11} varies with the avc voltage across C_6. Therefore, the voltage across R_{11}, and in turn the bias voltage for Q_3, is affected by the avc voltage. Thus, two transistors, Q_2 and Q_3, work together to maintain a fixed output level regardless of the variations in magnitude of the applied rf. Avc is accomplished through the feedback circuit using the rf that was originally converted to a specific negative dc voltage for this purpose.

Chapter 11

SCRs and Other Devices

Bipolar junction transistors conduct substantial quantities of collector current when the base voltage is about 0.7 volt greater than the emitter voltage. Only then is the transistor turned on. Collector current drops radically when the base-emitter voltage is less than 0.7 volt. The transistor is considered as off when the applied base voltage is equal to the emitter voltage or is negative with respect to the voltage at the emitter.

Given the above facts, the bipolar transistor can be used as a switch. It is turned on when the base is biased 0.7 volt above the emitter voltage and turned off by putting the base voltage at zero or negative with respect to the emitter. If a small bulb is in the collector circuit, it can be turned on and off by simply applying different voltages to the base of the device.

The above data applies, of course, to the npn transistor. Reverse polarities must be applied to the terminals of the pnp device to achieve the same results.

In an n-channel junction FET, if the gate voltage is zero with respect to the source, drain current flows. To stop this flow of current, a relatively large negative voltage must be applied to the gate with respect to the source to turn it off. This voltage must be equal to or greater than the pinch-off voltage. Voltages with the reverse polarity must be applied if a pnp device is being used for this purpose.

You just saw two examples of transistors used as switches. A number of other electronic switching devices have been developed with characteristics which differ somewhat from those of transistor switches. Of these, the popular ones are the silicon controlled rectifier or SCR and the triac. The devices used in conjunction with these switches are the unijunction transistor or UJT, the programmable unijunction transistor or PUT, and the diac. These latter devices can also perform some types of switching action.

SILICON CONTROLLED RECTIFIER

The SCR is essentially a diode that does not conduct in either direction when a low voltage is applied to it unless a pulse is placed for an instant at a third terminal of the device. This third terminal is known as the *gate*. The schematic representation of the SCR appears in Fig. 11-1.

Should the anode of an ordinary diode be made slightly positive (just somewhat above +0.4 volt) with respect to the cathode, it would not hesitate to conduct. The SCR does not conduct unless that applied voltage is quite large. Among other factors, SCRs are rated by the amount of voltage that must be applied to the diode terminals for it to start conducting. This is known as the *forward breakover voltage*, $V_{(BO)}$. While this may be as low as 50 volts in some cases, it can range as high as 1500, 2000, or even more, volts, for selected SCRs.

Should negative voltage be applied to the anode with respect to the cathode, a small quantity of leakage current flows. It does not get very large until a voltage, V_{RRM}, is exceeded. This is referred to as the *peak inverse voltage*. This voltage is usually close in magnitude to that of the forward breakover voltage. Once V_{RRM} is exceeded, the reverse current increases rapidly, but the voltage across the SCR rises very slowly above V_{RRM}. This is similar to the behavior of a zener diode. As for the gate voltage or current, it has no effect on V_{RRM} or on the voltage at which the SCR just

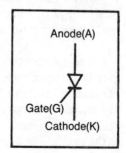

Fig. 11-1. Schematic representation of an SCR.

begins to conduct in the reverse direction.

Getting back to $V_{(BO)}$, any attempt to put a higher forward voltage across the diode will be resisted by the device. Assume it is in the circuit shown in Fig. 11-2. When the applied dc voltage is less than $V_{(BO)}$, the SCR does not conduct. It acts as if it were an open circuit. As soon as the applied voltage exceeds $V_{(BO)}$, the diode conducts. The amount of current it conducts at the instant $V_{(BO)}$ is exceeded is I_H, the *holding current*. Voltage across the diode drops to about 0.7 volt as soon as it starts to conduct. Now the bulk of the applied voltage is across R. From here on, the SCR behaves as if it were an ordinary diode. As more voltage is applied to the circuit, the current keeps rising rapidly, while the voltage across the device increases slowly above 0.7 volt. Essentially, it follows the voltage/current pattern of a junction diode.

Should the applied voltage be reduced while the SCR is conducting, current drops. But the SCR keeps conducting even as the current is reduced. It stops conducting when this current falls below I_H. Now the applied voltage must be greater than $V_{(BO)}$ for the diode to once again start conducting. That is, it must be above $V_{(BO)}$ only if a voltage or current pulse is not applied to the gate.

The gate forms a diode with the cathode of the SCR. To turn on the SCR, a positive voltage is applied to the gate with respect to the cathode through a resistor. The resistor is in the circuit to limit gate current to a safe value. Circuits in series with the anode and cathode of the SCR will conduct current even when the applied dc between the SCR terminals is as low as 0.7 volt. It will conduct under these conditions only if a voltage or current pulse has been applied to the gate. Conduction starts if the gate-to-cathode voltage is greater than V_{GD} or the gate current is greater than I_{GD}. Information about the amount of instantaneous gate voltage and current needed to turn on the SCR is supplied by the manufacturer of the device being used.

Only a short pulse is required in the gate circuit to turn on the SCR. The pulse can then be removed, but the SCR keeps conducting. It conducts until after the anode-cathode current drops below I_H.

A complementary SCR is also being manufactured. Here, the gate is at the cathode as shown in the schematic. The schematic of the complementary SCR is identical to that of the PUT discussed below, and is shown in Fig. 11-9. So if you see a diagram such as the one shown here, you will be able to understand its operation in the circuit. The behavior of the complementary and regular SCR

310

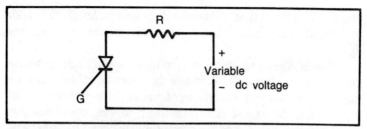

Fig. 11-2. Voltage applied to the anode and cathode of the SCR.

is the same, with the exception of the relative polarities of the voltages applied to their gates.

DIACS

Schematically, a diac looks like two diodes connected in parallel, with the cathode of one device connected to the anode of the other. This can be seen in Fig. 11-3. The diac does not conduct unless a specific minimum voltage is applied to its terminals (through some sort of resistance, of course). Once it conducts, a large current will flow. Current through the diac increases rapidly, while voltage across the device drops. Therefore, a large voltage will cause a large current pulse to pass through the device followed by a drop in voltage between terminals MT1 and MT2. Note that this action applies regardless of the relative polarity of the voltages applied. As a result, either ac or dc can be used to initiate conduction.

TRIACS

The SCR is essentially a diode that can be switched on and off by applying a voltage or current pulse to an extra terminal—the gate. But the SCR conducts current in only one direction. It can be used as a switch for dc. For ac pulses, it can be used as a switch during the portion of the cycle when the applied voltage is positive at the anode with respect to the cathode. When the SCR is off, a small amount of current flows. When it is on, electron current flows from the cathode to the anode and conventional current flows from

Fig. 11-3. Schematic symbol of a diac.

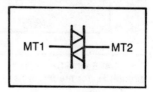

the anode to the cathode. It passes dc or half cycles of ac. In addition, it turns off after each half cycle because its current drops below I_H.

Two SCRs can be placed in a circuit to conduct current for the full ac cycle as in Fig. 11-4. When the upper ac voltage terminal is positive with respect to the lower terminal, current flows through R_1 and D_1 to the gate of SCR-1, turning it on. Because of this, current flows through R_L and the anode-cathode circuit of SCR-1 during this half-cycle. During the alternate half-cycle, current flows through R_2, D_2 and the gate-cathode circuit of SCR-2. SCR-2 is turned on to conduct additional current through R_L. Now the current flows through the second half-cycle. Thus, substantial current flows through R_L during both halves of the cycle. A full cycle of the ac voltage is developed across this resistor.

Instead of using two SCRs, one triac can be put into this type of circuit. A *triac* is essentially a diac with a gate. The schematic representation and an application similar to the dual-SCR circuit is shown in Fig. 11-5.

The triac turns on regardless of the polarity of the voltage applied to the gate and regardless of the polarity of MT2 with respect to MT1. Considering the various polarity situations, the smallest quantity of gate current is required to turn on the triac when the gate and MT2 are the same polarity with respect to MT1.

Just as with the SCR, in the absence of a gate pulse the triac will require a large voltage, $V_{(BO)}$, to turn on in either direction. Once turned on, it assumes the characteristics of a diode. Voltage across the device drops to about 0.7 volt and the current rises rapidly. While a voltage with the proper polarity is required to turn on the SCR, voltages with either polarity at the MT1 and MT2 ter-

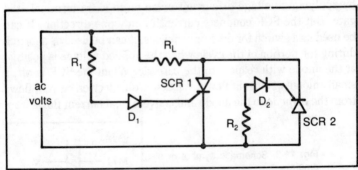

Fig. 11-4. Each SCR in the circuit turns on for one-half cycle. Current flows through R_L for the full cycle.

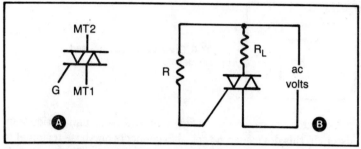

Fig. 11-5. The triac. A. Schematic representation. B. Circuit to conduct ac current for a full cycle.

minals of the triac will cause it to turn on. It stops conducting when the MT2 - MT1 current drops below I_H, the holding current. As with the SCR, a voltage or current pulse at the gate of the triac will turn it on. It will be turned on despite a low voltage across the MT2 to MT1 terminals.

When the upper terminal of the ac supply in Fig. 11-5B is positive with respect to the lower terminal, current is applied to the gate of the triac through R and the triac turns on. One positive half-cycle of current flows through the triac and R_L. Due to this current, a positive half-cycle of voltage is developed across this resistor. When the current drops below I_H, the triac stops conducting until the upper terminal of the ac supply is negative with respect to its lower terminal. Now a negative pulse is applied to the gate through R, turning on the triac. A negative half-cycle of current flows through R_L so that a negative half-cycle of voltage is developed across this resistor. As before, the triac turns off when the current in the device drops below I_H. The conducting sequence is repeated for all subsequent cycles of ac voltage that are applied to the circuit.

Positive and negative half-cycles of current flow through R_L. Thus, the voltage developed across this resistor takes the shape of the full cycle of ac voltage with the exception of a slight loss near zero. This missing section of the waveform occurs at the interval before the current applied to the gate is large enough to trigger the triac. For all practical purposes, you can say that the triac conducts for a full half-cycle while an SCR in a similar circuit would conduct only on alternate half-cycles.

UNIJUNCTION TRANSISTOR

A special device that can be used as the basic component in

313

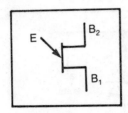

Fig. 11-6. Drawing of a UJT as used in schematics.

an oscillator or as a device to trigger an SCR is the unijunction transistor. In this device, an n-type slab connects one base terminal to a second base terminal. The terminals are assigned the symbols B_1 and B_2. The emitter terminal, E, is on a p-type material. The p-type material forms a junction at one point along the length of the n-slab. This is all shown in Fig. 11-6. Do not confuse this device with an FET. The schematics differ in the angle at which the third terminal is shown connecting B_1 to B_2.

The n-type slab can be considered as consisting of two resistances, R_{B1} and R_{B2} as in Fig. 11-7. One resistance is from the junction of the two semiconductors to B_2, and the second is the resistance from the junction to B_1. If E_{BB} volts is applied from B_2 to B_1, the voltage across R_{B1} is $[R_{B1}/(R_{B1} + R_{B2})]E_{BB}$. This is determined from the voltage divider equation. The resistance ratio in the equation is known as the *intrinsic stand-off ratio* or η, the Greek letter eta. The voltage at the junction of the semiconductors due to E_{BB} is therefore ηE_{BB}. The intrinsic stand-off ratio is an important characteristic of the UJT. It indicates the portion of the applied voltage that must be between the junction and B_1 if the UJT is to conduct. Because of the voltage developed across the forward-biased junction, the UJT will conduct when the voltage between E and B_1 is about 0.5 volt more than ηE_{BB}.

UJTs are frequently used in relaxation oscillator circuits. One such arrangement is in Fig. 11-8. When E_{BB} is initially applied to the circuit, C does not charge and behaves as if it were a short cir-

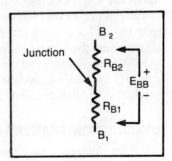

Fig. 11-7. The n-type slab of the UJT is split into two resistances.

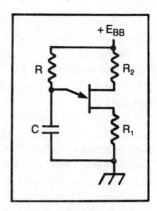

Fig. 11-8. UJT in a relaxation oscillator circuit.

cuit. But as time passes, it charges and voltage develops across C. When that voltage exceeds $0.5 + \eta E_{BB}$, the UJT is turned on. C discharges through the circuit formed by the E-B1 junction, R_{B1} and R_1. The charging and discharging cycle keeps repeating. Because of this, cyclic voltages are developed across all resistors in the circuit as well as across the capacitor. The circuit is in a state of oscillation. The frequency of oscillation is about equal to 1/RC.

THE PUT

The programmable unijunction transistor or PUT is used the same way as the UJT. The schematic of this device and a typical circuit are shown in Fig. 11-9. It has three terminals—the anode, cathode, and gate (A, K, and G).

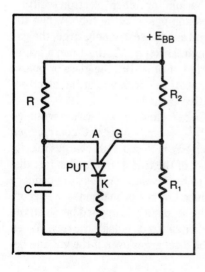

Fig. 11-9. The PUT oscillator.

Unlike the UJT, the η of this circuit is set by two resistors outside of the device, R_2, and R_1 and is equal to $R_1/(R_1 + R_2)$.

In Fig. 11-9, R and C are identical with R and C in Fig. 11-8. Thus, the anode of the PUT is used as the equivalent of the emitter of the UJT, while the cathode is used as the equivalent of UJT terminal B_2.

It should be noted that neither the UJT nor PUT must be used as oscillators. If the designer so chooses, he can use these devices as time delay switches. For example, if a short is placed across C, neither device will conduct. The devices in both circuits will conduct only after the short has been removed and a period of time has elapsed. The amount of time that elapses between the removal of the short and the start of conduction depends upon the RC network and the η of the UJT. In the PUT circuit this latter factor is determined by the resistors R_1 and R_2.

ELECTRONIC FLASHTUBES

Stroboscopes make use of electronic flashtubes. The circuits are designed to keep these tubes flashing at a steady pace and frequently involve the use of SCRs. Flashtubes are also used in high-speed photographic applications where short exposure times are desired.

The flashtube consists of three elements, as illustrated in Fig. 11-10. The anode and cathode are in a glass or quartz tube enclosure. The anode is usually made of pure tungsten, while the cathode usually contains some barium or other electron-emitting material. A small amount of xenon gas is in the tube. When the voltage between the cathode and anode is extremely large, the gas ionizes and current flows through the xenon. At this time, a bright light is emitted for a short period of time. But the glass or quartz remains relatively clean because of the presence of the electron-emitting material in the negative cathode.

A trigger electrode consisting of a few turns of wire is wrapped around the tube. When switch SW is open, capacitor C gets charged slowly through resistor R. When SW is closed, C discharges rapidly through the switch and primary of the transformer. This fast discharge causes a high voltage to be developed across the primary winding. A high voltage is induced by the primary into the secondary. The high voltage from the secondary is applied between the cathode in the tube and the trigger electrode outside the tube. When this voltage is present, the tube discharges even if the voltage between the anode and cathode is insufficient to cause the breakdown by itself.

Anode to cathode voltage is usually double the peak voltage from the line. Thus, a doubler is shown as supplying this voltage—about 340 volts dc—in Fig. 11-10. But this voltage by itself is insufficient to cause breakdown. A voltage pulse between 1000 and 3000 volts is applied to the trigger element through the transformer. At this instant, conduction takes place in the tube and the tube flashes at the moment the pulse is applied.

APPLICATIONS

Kits are available using many circuits similar to those described here. These kits are supplied by Eico, Heath, Radio Shack, and so on. It is interesting and enlightening to build these as well as other types of educational kits. The descriptions supplied here will give you a general insight as to how the components are used in circuits. The kit manuals will provide more detailed information about the actual item you are building.

Burglar Alarm

The sensing switches used at windows in burglar alarm setups are often normally open circuits. That means that the switch completes a circuit or closes when the window is opened. In other cases, normally closed switches are used. An alarm circuit is tripped when the switch is opened due to an unlawful opening of a closed window. A circuit which reacts to both types of switching circuits, is shown in Fig. 11-11.

Switch SW1 is normally open. No current flows through the

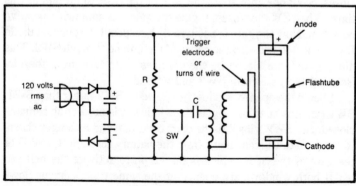

Fig. 11-10. Flashtube in a circuit.

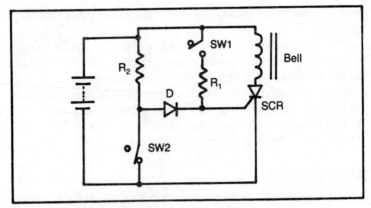

Fig. 11-11. Burglar alarm circuit.

switch to the gate of the SCR to supply it with current. Switch SW2 is normally closed. It connects the lower terminal of R_1 to ground so current from that resistor cannot flow though diode D to the gate of the SCR. It remains off when SW1 is open and SW2 is closed.

Now assume that the normally open switch is used at one window in a house and that the normally closed switch is at the second window. Start by considering the window with the normally open switch. If that window is opened, SW1 closes. Current flows from the battery through SW1 and R_1, to the gate of the SCR. It is turned on due to this gate current. When the SCR conducts, current flows through it and through the bell in its anode circuit. The bell then rings and sounds the alarm.

During this time, SW2 just stays closed as if it were a short circuit. It is not activated at any time when the window it is monitoring remains closed. Even through SW2 remains closed when SW1 closes, SW2 does not short the gate of the SCR to ground at this time. When SW1 is closed, it puts the gate at a somewhat positive potential. Now the gate is positive with respect to ground. A diode D is connected between the gate and the ground through SW2. This diode is reverse biased by the voltages at its terminals, thereby isolating the gate from SW2 and ground.

Should the opened window be served by SW2 rather than SW1, SW1 remains open so long as the window it is protecting remains closed. But SW2 trips into the open condition and no longer shorts R_2 to ground. Current flows from the battery through R_2 and D to the gate of the SCR, which is turned on to activate the bell.

If both windows are opened at the same time, current flows through R_1 and R_2 to the gate of the SCR, turning it on.

Audio-Modulated Lights

In Fig. 11-12, audio from an amplifier is applied to the primary of a transformer. From there, it is fed through R_2 to the gate of the SCR. Voltage from the audio signal keeps tripping the SCR. When it conducts, current flows through bulb B_1, but the SCR turns off when the negative half-cycle is applied to its anode-cathode circuit. The SCR is turned on again by the next audio pulse.

Should audio be available on a continuous basis, it keeps tripping the SCR so B_1 lights during each alternate half-cycle. The maximum light is available from the bulb. If the audio comes in short and infrequent pulses as is normally supplied by music and speech, the SCR is tripped at a random rate. The number of times the SCR is tripped during any period of time depends upon the number of pulses available during this time. Thus, the light will turn on more often when longer and more frequent audio pulses are available than when these pulses are of short duration and are present at relatively infrequent intervals. If the gaps between the pulses are short, the light appears to be on continuously, but the brightness will be proportional to the number of times the SCR is activated in a given period of time.

In the absence of audible pulses, the SCR will remain off and B_1 will not light. Variable resistor (potentiometer) R_1 is used to adjust the magnitude of pulses applied to the SCR. With the control in its maximum setting, audio from the transformer is not attenuated. In this case, lower voltage pulses may be of sufficient magnitude to trip the SCR. Because both low-and high-voltage pulses are readily available, pulses of both magnitudes will be adequate to trip the SCR. With this input, the SCR will be on during all half-cycles. B_1 will appear as if it is on continuously so its brightness will be at a maximum.

Should the level control be set so that its wiper is near the bottom of the resistor, all audio from the transformer is attenuated. The low voltage pulses may be attenuated to the level where they

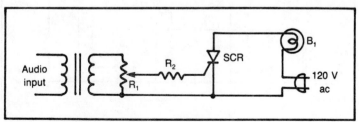

Fig. 11-12. Lights are activated by the audio input.

are insufficient to trip the SCR. But the high voltage pulses are still capable of making the SCR conduct, so the bulb will light only when high-voltage pulses are present. The on periods are now less than when the control was at maximum because fewer pulses are available to turn on the bulb. The brightness of the bulb fluctuates more readily than before because the activating pulses are now present on a more random basis.

Three-Channel Sound/Light Converter

The sound-to-light converter in Fig. 11-13 is basically a combination of three circuits, each similar to the one in Fig. 11-12. Voltage is applied to the gates of the SCRs from audio at the secondary of the transformer. But despite the fact that the overall audio signal is at the secondary of the transformer, only a portion of the band is available to trip the gates of the different SCRs.

SCR1 is turned on by signal voltages present at all portions of the audio band. A good portion of the low frequencies are eliminated by the R_1 - C_1 filter before the audio is fed to the gate of SCR2, so signals at the middle and high-frequency portions of the audio band are applied to this gate. Only high frequencies are available at the gate of SCR3 to turn it on because of the characteristics of the R_2-C_2 filter.

Different color bulbs are in the anode/cathode circuits of the various SCRs. As shown in the figure, a blue light goes on each time SCR1 is activated, a yellow light each time SCR2 is activated, and a red light when SCR3 is turned on. Thus, only high frequencies present in the audio signal turn on the red bulb, the middle

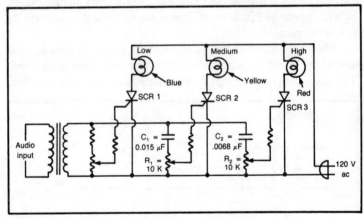

Fig. 11-13. Three channel sound-to-light converter.

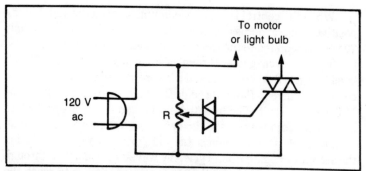

Fig. 11-14. Motor speed control or light dimmer. Conduction is varied from a period of 180° to 90° during each half-cycle.

and high frequencies both turn on the yellow bulb, and any frequencies present in the audio signal turn on the blue light. The amplitude of the signals used to activate each SCR is controlled by a potentiometer in its gate circuit. When the controls are adjusted properly, the blue light goes on due to the low frequencies in the audible signal, the red light goes on only when high frequencies are present, and the yellow light glows only when frequencies in the middle portion of the audio band are present. This is the basis of the "color organ."

Conduction-Time Limiter

A light dimmer or motor-speed control circuit is shown in Fig. 11-14. Current is applied through the diac to the gate of the triac. No current flows through the diac until its breakdown voltage is exceeded. The voltage across the diac drops rapidly as the current passes through it to the gate of the triac. The triac is turned on and current then flows from the 120-volt ac line through triac terminals MT1 and MT2 as well as through the motor or bulb in the circuit.

It is easy to understand the action of the circuit if you assume that 20 volts is needed at the wiper of variable resistor R to activate the triac during each half-cycle. When R is set to maximum, the maximum voltage is applied to the gating circuit from the 120-volt supply. The ac input voltage passes through its 360° cycle. It starts at the 0-volt level at 0°, rises to a positive peak of 120 volts × 1.41 or about 170 volts at 90°, and then drops back down to 0 at 180°. During the second half of the cycle, the voltage sequence repeats but is negative, dropping to −170 volts at 270° and rising back to 0 at 360°.

When the control is set to maximum, it was indicated that the positive peak ac across this potentiometer is 170 volts. But only 20 volts is required to turn on the diac, and, in turn, the triac. Thus, the first half-cycle passes through a small angle before the 20-volt level is reached and the diac and triac are turned on. Once the triac conducts, current flows through the bulb or motor in the triac MT1/MT2 circuit. The triac turns off when the ac drops back to nearly zero because now the current through the device is less than I_H. The triac is on for almost the full positive half-cycle. During the second or negative half-cycle, the waveform must pass through a similar small angle beyond 180° for the applied ac to reach the −20-volt level. Once this level is reached, the triac turns on for just about another full half-cycle. Thus, current flows through the motor or bulb for just about a complete cycle. Motor speed or light intensity is at a maximum.

Should the wiper on the control be put at a near-minimum setting so that the peak voltage at the wiper is 20 volts rather than 170 volts, current will pass through the diac to the gate of the triac only when the supply voltage is at its positive or negative peak. During the positive half-cycle, the triac will not start to conduct until the cycle is at the 90° point, because only then is the voltage at the input to the diac and gating circuit enough to turn on the triac. Under these conditions, current flows through the triac and motor or bulb from the instant the supply is at its peak at 90° until it reaches the 0 volt level at 180° during the positive half-cycle. Because the supply voltage is zero at 180°, current drops below I_H and the triac turns off. Similarly, current flows in the negative half-cycle during the 270° to 360° portion of the cycle. Because current flows only half of the time, the average current reaching the motor or bulb is now less than when current flowed through almost the entire two half-cycles. As a result, the motor runs more slowly, or the quantity of light supplied by the bulb is less than when the wiper was set at maximum.

The situations at the two extreme settings of the control have been described. At intermediate settings, the brightness of the bulb or the speed of the motor is somewhere between these two extremes.

You just saw that only a 90° variation of starting times exists during each half-cycle if the circuit in Fig. 11-14 is used. This range can be extended to an almost 180° variation in each half-cycle by using the two circuits in Fig. 11-15. This extension of the range is accomplished through use of an RC phase-shift network. Volt-

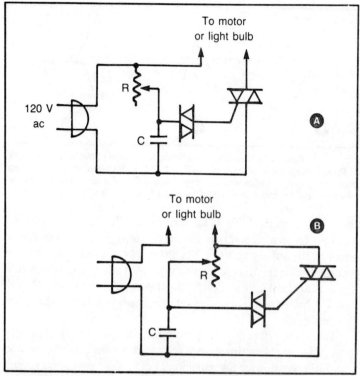

Fig. 11-15. Motor speed control or light dimmer using a phase-shift network. A. Applying voltages to the triac gate through a diac. The voltages can be phase shifted to 90°. B. Adding feedback between terminal MT2 of the triac and its gate.

age developed across C is applied to the gate-MT1 circuit of the triac through the diac. This affects the turn-on time in the two half-cycles. Because the same factors apply to both half-cycles, only details concerning the positive half-cycle will be described here.

A phasor diagram of the RC network is shown in Fig. 11-16A. The voltage across capacitor C lags the circuit current and hence the voltage across R by 90°. But the sum of the voltages across R and C must be equal to 120-volts ac. This is shown in Fig. 11-16A and was described in a previous chapter. Note the phase angle, α, between the supply voltage and the voltage across the capacitor.

When the wiper on the control makes contact with the top of the resistor in Fig. 11-15A, R is equal to 0 ohms. When this occurs, the entire supply voltage is across C and α is 0°. The voltage across C is at the gating circuit of the triac. Now, it is in phase with

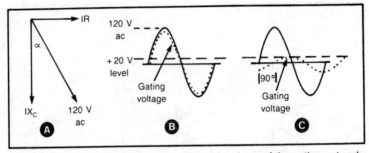

Fig. 11-16. Illustrating the effects of shifting the phase of the gating network. A. Phasor diagram of RC network in Fig. 11-15. B. Relative amplitudes of applied and gating voltages when R is very small. C. Relative phase and magnitudes of applied and gating voltages when R is very large.

the voltage at its MT1-MT2 terminals. This is shown in Fig. 11-16B. The voltage across the capacitor, the gating voltage, is equal to the supply voltage. The 20 volts required to trigger the diac and cause the triac to turn on is present when the applied voltage is near the 0° point. Conduction takes place for nearly the full positive half-cycle. This is, of course, repeated during the negative half-cycle.

When the wiper is near the bottom of the control, R is very large. Almost the entire supply voltage is across this resistor and very little remains for the capacitor. Let us say that the maximum voltage across the capacitor under these conditions is 20 volts. When this occurs, α is close to 90°. This is shown in Fig. 11-16C. With the small voltage across the capacitor, the triac will be triggered when the voltage is at its 20-volt peak. At this time, the cycle has almost passed through its full 180°. The triac will stop conducting soon after it starts. The triac will conduct for a very short portion of the cycle. It ceases to conduct when the ac voltage applied to its MT1-MT2 circuit drops to near zero volts and the current in that circuit is reduced below I_H. Under these conditions, almost no current flows through the motor or bulb, so the device is either off or is operating at minimum capacity.

In one setting of control, you find that the triac conducts for almost 180° in each half-cycle. In the second setting, it conducts for close to 0° in each half-cycle. Thus, the conduction period will be varied over a 180° range during each half-cycle in different settings of the control. By varying the conducting period, the speed of the motor in the circuit (or the brightness of the bulb) also varies. The control can therefore be used to adjust the motor speed or light intensity to the desired level.

The circuits in Fig. 11-15A and B behave in a similar way, except that the one in B has feedback added from MT2 to the gate of the triac. No such feedback exists in the circuit in A because the RC circuit there is connected directly across the power line.

Battery-Operated Light Flasher

In Fig. 11-17, a UJT is wired as a low-frequency oscillator. Its frequency varies with the setting of the variable resistor, R. R and C are usually specified for the UJT to provide about one pulse every second when R is at its maximum to about five pulses each second when R is at its minimum. These pulses are developed across resistor R_1. This resistor is wired in the Base one (B_1) circuit of the device.

Pulses are applied from B_1 to the gates of SCR1 and SCR2. The first one in a series of pulses turns on both SCRs. The bulb in the drain circuit of SCR1 lights and remains lit. As for SCR2, the resistor R_4 in the anode circuit is relatively large. Therefore, current through SCR2 is limited to less than I_H. As a result, it turns on for an instant and turns off within a short portion of a second. But SCR1 remains on.

When neither SCR is on, voltages at the anodes of the two SCRs are identical at +12 volts. At this time, identical voltages are at both ends of capacitor C_2. But this changes when SCR1 is on. The voltage at its anode drops to almost one volt. SCR2 is off so its anode is at +12 volts. Now C_2 gets charged.

Plus 12 volts is at one end of C_2. The anode of SCR1 is on so its anode is at about one volt. This is at the other end of C_2. Dur-

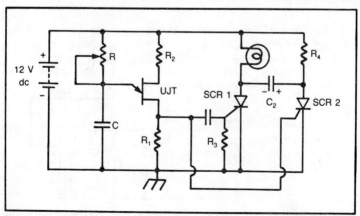

Fig. 11-17. Light flasher.

'ng the interval when SCR1 is on, C_2 gets charged to $12 - 1 = 11$ volts with the polarity shown in the drawing.

At the next pulse from the UJT, SCR2 is turned on for an instant. While it is on, the anode voltage drops. Charged capacitor C_2 puts the anode of SCR1 at about 11 volts below that of the anode of SCR2. Because the anode voltage of SCR1 is now less than its cathode voltage, the current in SCR1 drops below I_H, which stops conducting, and the bulb is turned off.

This process keeps repeating so that every other pulse from the UJT activates SCR1 to turn on the bulb, while the other pulses activate SCR2 so that the bulb gets turned off. The on and off or flashing frequency for the bulb depends upon the frequency with which the UJT supplies the triggering pulses to the gates of the SCRs.

A PUT can be substituted for the UJT to supply the pulses to operate the circuit. The intrinsic stand-off ratio can now be adjusted to optimize the frequency of the flashes.

Strobe Light

A voltage-doubler circuit supplies about 340 volts dc to the + and − (or anode and cathode) terminals of the flashtube in the circuit drawn in Fig. 11-18. This voltage is also applied to the anode of the SCR through R_1. The 340 volts is reduced by the divider circuit formed by R_2 and R_3 and applied to the gate of the SCR. The SCR is turned on when there is enough voltage at the gate. While the SCR is off, capacitor C_4 is charged through the trans-

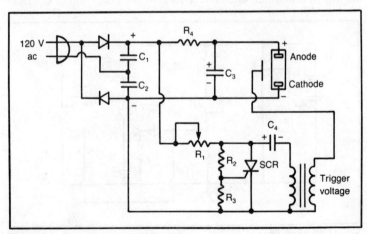

Fig. 11-18. Variable-speed strobe light.

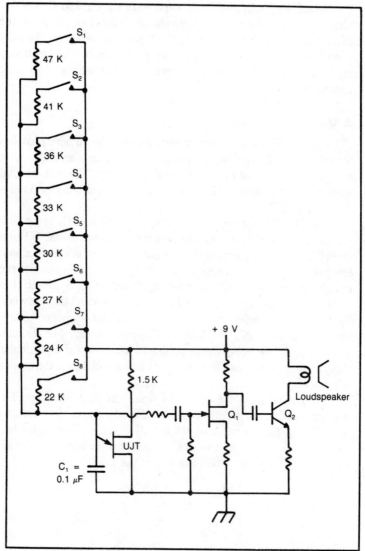

Fig. 11-19. A musical instrument.

former. A pulse from C_4 flows through the SCR when it is turned on. The pulse also flows through the primary of the transformer through which C_4 was previously charged. The voltage due to this pulse is induced into the secondary winding of the transformer and magnified because of its turns ratio. A large voltage is applied from this secondary to the trigger winding of the flashtube and it lights.

When the flashtube lights, capacitors C_1, C_2, and C_3 are discharged for an instant. This low voltage pushes the current in the SCR below I_H, so it turns off. As the capacitors recharge, the SCR is once again turned on to trigger the second flash. The frequency of these repetitive flashes is determined by the setting of variable control R_1.

A Musical Instrument

As with all musical instruments, the one shown here provides tones of different frequencies. Figure 11-19 shows an instrument that is often referred to as an "electronic organ." How it got that name is anyone's guess. Here, however, only eight individual tones are provided.

The UJT in this circuit is used as an oscillator. C_1 is the first component that determines the frequency of oscillation. When a switch is closed, the second component, a resistor, is switched into the frequency-determining circuit. The resistor that is put into the circuit depends upon which switch is closed. With S_1 closed, the oscillation frequency is about 262 Hz, while it is about 523 Hz when S_8 is closed. Between these two extremes, the sequence of frequencies is about 294 Hz, 330 Hz, 350 Hz, 392 Hz, 440 Hz, and 494 Hz, each one being the appropriate frequency of a particular musical note. A tune can be played by closing switches in the proper sequence.

The output from the UJT is increased in magnitude by FET voltage amplifier Q_1 and power amplifier Q_2. An FET is used as Q_1 so that its high impedance will not affect the UJT oscillator frequency, while the output from Q_2 is fed to a loudspeaker. As always, a PUT can be used rather than the UJT.

Chapter 12

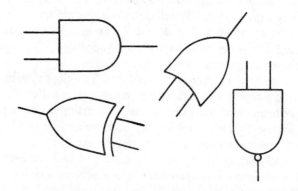

Digital Electronics

If the measurements of some quantity can vary continuously up or down the scale, then this is *analog* data. Thus, measurements of some factor such as current, voltage, or resistance are analog data, and these factors are known as *analog variables*. The measurements vary up and down with physical quantity present in the real world. It is a directly measurable quantity. Voltage amplifier circuits which provide output voltage with magnitudes that are related to the magnitudes of the voltages at the input are analog amplifiers.

The digital representation of a quantity presents data in the form of discrete steps or counting units. Counting may be accomplished by noting a number of fingers, toes, or electrical pulses.

Regardless of the voltage at the input of a digital gating circuit, the output is either a specific maximum voltage or a specific minimum voltage with no variations in between. Special digital circuits may be designed in which a series of identical high and low voltages are at the output. This output is a code formed by steps. So if a digital circuit is to measure a voltage, the output is in the form of discrete numbers that are represented by a quantity of steps. The voltage in a digital measuring circuit is not read on a meter scale, because here the pointer can indicate any discrete number or any intermediate value in between two numbers. Digital readouts are available only as discrete or whole numbers.

NUMBER SYSTEMS

Different number systems are used to implement digital processes. All of these systems may be considered as codes. In the numbers we use in everyday calculations, there are ten quantities in the code, namely 0, 1, 2, 3, 4, 5, 6, 7, 8, and 9. You can specify any magnitude or perform any arithmetic operation when you understand the meaning of each of these digits.

Here we have a system with ten digits. Two-digit systems are used in most digital work although eight- and sixteen-digit systems have also been found to be very useful. Converting from one numbering system to the other is an important procedure in all types of digital work but especially in computer applications.

Decimal and binary systems are described in detail here and should be understood if digital work is to be accomplished successfully. There are also brief descriptions of octal and hexidecimal systems. To understand these completely, the material presented must be read slowly and a concerted effort made to apply the methods shown. The latter two numbering systems are not elaborated upon further because they are not necessary to understanding and using basic digital circuit techniques. A working knowledge of these systems is essential if you are dealing with computer electronics.

The decimal system will be discussed first. This is the code we use when performing our daily tasks.

Decimal Numbers

In this system, every number is based on ten digits, 0 through 9. Decimal numbers above 9 are combinations of two or more of these ten digits. Thus, number 10 is composed of a 1 and a 0 to form a 10, a 36 is composed of a 3 and a 6 to form a 36, and so on. The position of a digit in the overall number gives it a particular significance. If the number is, for example, 5713, the 5 is the *most significant digit* or MSD because it contributes the largest quantity to the magnitude of the number. In this number, 3 is the *least significant digit* or LSD because it contributes the smallest quantity to the magnitude of the number. The significance of the individual digits increases as you progress from the LSD to the MSD. The 1 is more significant than the 3 in the number 5713; the 7 is more significant than the 1; and the 5 is more significant than the 7.

You can determine the exact significance of each digit by breaking the number down into a sum. This quantity is equal to the number of digits comprising the number. Thus, 5713 can be regarded as the sum of four numbers, namely

$$5000$$
$$700$$
$$10$$
$$3$$

The sum of these numbers is 5713.

Each number can be represented as a digit multiplied by the base 10 to some exponent. Thus,

$$5000 = 5 \times 10^3$$
$$700 = 7 \times 10^2$$
$$10 = 1 \times 10^1$$
$$3 = 3 \times 10^0$$

Because this is a decimal system with 10 digits, 10 is the base. The LSD in this number is the digit in that position of the number multiplied by 10^0. The next significant digit is multiplied by 10^1 and so on until you multiply the MSD by 10 to the exponent for the MSD. Here it is 10^3. So if you want to represent these digits in order of significance, you can write them as

$$5 \times 10^3 + 7 \times 10^2 + 1 \times 10^1 + 3 \times 10^0$$

Copying each digit in the sequence shown gives 5713.

In order to identify the system in which a particular number belongs, a subscript indicating the base is placed after the number. Here, this subscript is 10. To show that the number 5713 is in the decimal system, it is written as 5713_{10}.

Binary Numbers

Digital electronics is based on a two-number system. These numbers are 0 and 1. In this system, digits are referred to as *bits*. Because there are only two bits in the system, the base is 2.

If you should encounter a binary number such as 1101001, the significance of each bit is 2 to an exponent. The least significant bit at the right of the number has a value of that bit multiplied by 2^0 or $1 \times 2^0 = 1 \times 1 = 1$. The significance of the next bit to the left is $0 \times 2^1 = 0 \times 2 = 0$. Progressing from there to the MSB (most significant bit) at the left, the values represented by the various bits are $0 \times 2^2 = 0 \times 4 = 0$, $1 \times 2^3 = 1 \times 8 = 8$, $0 \times 2^4 = 0 \times 16 = 0$, $1 \times 2^5 = 1 \times 32 = 32$ and $1 \times 2^6 = 1 \times 64 = 64$. Listing these numbers in sequence from the MSB to the LSB gives

$$
\begin{array}{rl}
\text{MSB:} & 1 \times 2^6 = 1 \times 64 = 64 \\
& 1 \times 2^5 = 1 \times 32 = 32 \\
& 0 \times 2^4 = 0 \times 16 = 0 \\
& 1 \times 2^3 = 1 \times 8 = 8 \\
& 0 \times 2^2 = 0 \times 4 = 0 \\
& 0 \times 2^1 = 0 \times 2 = 0 \\
\text{LSB:} & 1 \times 2^0 = 1 \times 1 = 1
\end{array}
$$

for a total of 105 in decimal numbers. Thus, 1101001_2 in binary form (the subscript 2 indicates that this is a binary number) is equal to 105_{10} in the decimal system.

It is obviously quite simple to determine what a binary number is in our decimal system by simply multiplying the bit by 2 to the proper exponent and adding these quantities. As examples,

$101 = 1 \times 2^2 + 0 \times 2^1 + 1 \times 2^0 = 4 + 0 + 1 = 5$
$1111 = 1 \times 2^3 + 1 \times 2^2 + 1 \times 2^1 + 1 \times 2^0 = 8 + 4 + 2 + 1 = 15$
$10111 = 16 + 0 + 4 + 2 + 1 = 23$
$11101110 = 128 + 64 + 32 + 0 + 8 + 4 + 2 + 0 = 238$

and so on.

It is a little more complicated to convert from decimal to binary than from binary to decimal. Because there are two bits or digits in a binary number, you apply the number 2 to the process.

Start by writing the decimal number at the right end of the page. Divide the number by 2, then write the quotient to the left of the original number. If there is a remainder when you divided the original number by 2, write the remainder under this quotient. Otherwise write a 0 here. Next divide this quotient by 2, write the result of this operation as the next number to the left of that quotient, and write the remainder under the new quotient. The sequence of remainders is the binary equivalent of the original decimal number. An example will clarify this procedure.

Suppose you want to know the binary equivalent of the decimal number 27. Following the procedure, you write 27 at the right end of the page. Divide that number and the sequence of quotients by 2. Write the remainders under the quotients. You will find that the binary equivalent of 27_{10} is 11011_2.

Quotient: $0 \leftarrow 1 \leftarrow 3 \leftarrow 6 \leftarrow 13 \leftarrow \underset{2}{27} \leftarrow 27$

Remainders: 1   1   0   1   1

You can check the result by converting it back to decimal number form. It is equal to the sum of $1 \times 16 + 1 \times 8 + 0 \times 4 + 1 \times 2 + 1 \times 1 = 27$.

Note the procedure used in the example. First, 27 was divided by 2. It gave us a quotient of 13 with a remainder of 1. So the 13 was written in the next column to the left of the 27 and the remainder 1 was written just below it.

Next, 13 was divided by 2 to give a quotient of 6 and a remainder of 1. The 6 was written in the next column to the left of the 13 and the remainder 1 was written just below the 6.

Continue by dividing 6 by 2 to get a quotient of 3 and no remainder, or a remainder of 0. Write the 3 in the next column to the left of the 6 and write the remainder of 0 just below the 3.

After dividing the 3 by 2, write a 1 in the column just to the left of the 3 and the remainder of 1 just below the 1.

Finally, divide the 1 by 2. Write the quotient of 0 in the next column to the left of the 1 and write the remainder of 1 just below the 0.

The binary equivalent of the decimal number is the sequence of remainders obtained from left to right as shown above.

Let's try a few examples.

1. Convert to 15_{10} to binary form.

$0 \leftarrow 1 \leftarrow 3 \leftarrow 7 \leftarrow \underset{2}{15} \leftarrow 15$

1   1   1   1

so the binary equivalent of 15_{10} is 1111_2.

2. Convert 11_{10} to binary form.

$0 \leftarrow 1 \leftarrow 2 \leftarrow 5 \leftarrow \underset{2}{11} \leftarrow 11$

1   0   1   1

so the binary equivalent of 11_{10} is 1011_2.

3. Convert 41_{10} to binary form.

$$0 \leftarrow 1 \leftarrow 2 \leftarrow 5 \leftarrow 10 \leftarrow 20 \leftarrow \underset{2}{41} \leftarrow 41$$

$$1 \quad 0 \quad 1 \quad 0 \quad 0 \quad 1$$

so the binary equivalent of 41_{10} is 101001_2.

Different codes are often used to represent decimal numbers in binary form. The numbers formed using these codes are not actual binary equivalents of the decimal number. They are just useful conventions to represent a series of decimal digits.

Among these codes, the *binary coded decimal*, or BCD system, is used frequently. Here, each digit in the decimal number is represented by a four-bit binary number. Thus, the decimal number 1 is 0001, 3 is 0011, 9 is 1001, and so on. If there is more than one digit in the decimal number, the overall number is represented by a number of groups of four binary bits, each group being equal to the digit it represents in the decimal number. For example, consider the number 3941_{10}. It is converted to BCD coded form as follows.

Decimal Digits:	3	9	4	1
BCD Equivalent:	0011	1001	0100	0001

Thus, 3941_{10} is 0011 1001 0100 0001 in the BCD code. As you can see, the number in the BCD code is not the actual binary equivalent of the decimal number but is only a decimal number represented by a code.

Octal Numbers

Another number system used frequently in digital work involves the eight digits, 0, 1, 2, 3, 4, 5, 6, and 7. If you count only from 0 to 7, these digits represent the same numbers as they do in the decimal system. Above 7, more than one digit is used to form a number.

This system has a base of 8. The decimal equivalent of the octal LSD is that digit multiplied by 8^0. Progressing from there to the left of an octal number with more than one digit, the more significant digits have magnitudes equal to the digit in question multiplied by 8^1, 8^2, 8^3, 8^4, and so on. Thus, the octal number 3512_8 is the equivalent of $3 \times 8^3 + 5 \times 8^2 + 1 \times 8^1 + 2 \times 8^0$ in decimal numbers or $3 \times 512 + 5 \times 64 + 1 \times 8 + 2 \times 1 = 1866_{10}$.

To convert a decimal number to octal form you use a procedure similar to that used when converting from decimal to binary numbers. But now you divide the numbers by 8 instead of by 2. For example, convert 1094_{10} to octal form.

Quotients: $0 \leftarrow 2 \leftarrow 17 \leftarrow 136 \leftarrow \dfrac{1094}{8} \leftarrow 1094$

Remainders: 2   1   0   6

so that 1094_{10} is the equivalent of 2106_8. Note that you cannot use a calculator to find this equivalent number because numbers in the calculator assume the base is 10. You must divide the numbers using pencil and paper and then get the remainders.

To convert from octal to binary, write the binary equivalent of each octal digit using three binary bits. For example, find the binary equivalent of the octal number 2106_8.

Octal: 2   1   0   6
Binary: 010   001   000   110

This can be determined by using a binary number equal to each octal digit. Here 2_8 is 010_2, 1_8 is 001_2, 0_8 is 000_2, and 6_8 is 110_2. Thus, 2106_8 in octal numbers is identical to 010001000110_2 in binary numbers.

Converting from binary to octal numbers, involves splitting the binary number into groups of three bits and getting the octal equivalent of each group of bits. For example, the octal equivalent of binary number 1101011001111_2 is

Binary: 001   101   011   001   111
Octal: 1   5   3   1   7

15317_8. Note that two 0s were added in front of the MSB so that all groups of binary numbers will be composed of three bits.

Hexadecimal Numbers

Besides the systems discussed above, the hexadecimal number system is much used in computer work. Here, there are sixteen basic digits—0 through 9 as in the decimal system—in addition to six digits represented by the letters A, B, C, D, E and F. These letters are the equivalent of the decimal numbers 10, 11, 12, 13, 14 and 15, respectively. For magnitudes above F (or 15), more than one digit must be used to represent the hexadecimal number.

In the hexadecimal numbering system, the base is 16. The magnitude of the LSD is 16^0 in decimal code multiplied by the digit. Progressing from right to left (or from the LSD to the MSD), the values increase by the value of the hexadecimal digit multiplied by 16^1, 16^2, 16^3, and so on. Thus, the hexadecimal number $3C9_{16}$ is equal to $3 \times 16^2 + C \times 16^1 + 9 \times 16^0 = 768 + 192 + 09 = 969$ in decimal notation. (Note that C_{16} is the same as 12_{10} so that the middle digit in the hexadecimal number converted here is $12 \times 16 = 192$.)

Decimal numbers can be converted into hexadecimal form by dividing the number and quotient by 16 and using the remainders for the hexadecimal number. Thus, the hexadecimal equivalent of 983_{10} is

$$\text{Quotients:} \quad 0 \quad 3 \quad 61 \quad \frac{983}{16} \longleftarrow 983$$

$$\text{Remainders:} \quad 3 \quad 13 \quad 7$$

or 3D7 because D is the same as 13. Here too, the arithmetic must be done on paper to get the remainders.

When converting from a hexadecimal to a binary number, each hexadecimal digit is represented by four binary bits rather than the three bits used when converting from numbers in the octal system. Hexadecimal number 983_{16} in binary form is

Hexadecimal:	9	8	3
Binary:	1001	1000	0011

so that 983_{16} is equal to 100110000011_2.

To convert from a binary number to a hexadecimal number, split the binary number into groups of four bits. Then get the hexadecimal equivalent of each group. Thus, number 1101011001111_2 in hexadecimal form is $1ACF_{16}$ and is derived as follows:

Binary:	0001	1010	1100	1111
Hexadecimal:	1	10	12	15

Here A is the equivalent of 10, C of 12 and F of 15. Note that three 0s were added in front of the MSB so that all groups of binary numbers will be composed of four bits.

LOGIC GATES

Three basic circuits are used in all digital equipment. These

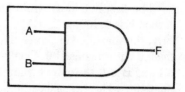

Fig. 12-1. Schematic of a two-input AND gate.

are referred to as *gates*. They are the AND gate, OR gate, and NOT gate. Every other digital circuit is a combination of these gates. In some cases, a single circuit is designed to perform as if it were constructed of a number of gates.

The outputs from the gates are controlled by only two voltages levels—a high and a low. If the voltage is above a certain level, it is usually assigned the number 1. Should it be below that level, the number assigned is 0. Because there are only two levels, only two binary numbers, 0 and 1, need to be used. All the different combinations of bits (or levels) applied to the inputs of a gate produce either a 0 or a 1 at the output.

AND GATES

Let us start our discussion with the AND gate. A diagram of this gate is shown in Fig. 12-1. As usual, in schematic representations of digital components, the power supplies are not shown, but it is obvious that power must be applied to the gate circuit if it is to operate.

A and B are the two input terminals to the AND gate and F is the output terminal. The characteristics of the AND gate are such that if A and B are both 1's, then F is at 1; if either or both A and B are 0, the F is also 0. This information is usually summarized in what is known as a *truth table*. A truth table for the AND gate is shown in Fig. 12-2. Here, the states of one input are shown in column A, and the states of the second input are shown in column B. Column F indicates the states of the output when the states of A and B are as indicated in their respective columns.

A	B	F
0	0	0
0	1	0
1	0	0
1	1	1

Fig. 12-2. Truth table for the AND gate in Fig. 12-1.

The first line in the truth table shows that if A is 0 and B is 0, then F is 0. In line 2, if A is 0 and B is 1, then F is still 0. Similarly, the third line indicates that if A is 1 and B is 0, F does not change but remains at 0. Finally in the last line, you see that if A is 1 and B is 1, the output from the AND gate is 1. This table substantiates what was said earlier, that the output from an AND gate is 1 only when inputs A and B are both 1.

You should note one other factor in the truth table. Inputs A and B are shown with all possible combinations of 1s and 0s. This is accomplished by using binary numbers for the sequence of data in the table. The first line for A and B is binary number 00, which is 0 in decimal notation. Binary number 01 is on the second line and is 1 in the decimal code. The third line shows the binary sequence 10 or 2 in decimal code. Finally, the fourth line has 11 for A and B. This binary number is the same as a 3 in the decimal number system. So the states, 0 or 1, listed for the inputs A and B, are in numerical order starting at 0.

AND gates can be constructed with considerably more than two inputs. In any case, all inputs must be a 1 if the output is to be a 1. An AND gate with three inputs in its truth table is shown in Fig. 12-3. Note the decimal magnitudes of the numbers in the sequence of input states shown in the truth table. A total of eight variations is possible here, so binary numbers with decimal values from 0 to 7 are shown to give all possible combinations of input states.

OR GATES

This gate differs from the AND gate in that at least one of the inputs must be 1 if the output is to be 1. It has the name OR be-

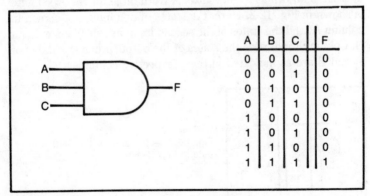

A	B	C	F
0	0	0	0
0	0	1	0
0	1	0	0
0	1	1	0
1	0	0	0
1	0	1	0
1	1	0	0
1	1	1	1

Fig. 12-3. Three-input AND gate with the truth table.

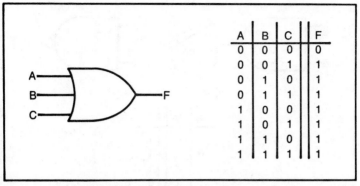

A	B	C	F
0	0	0	0
0	0	1	1
0	1	0	1
0	1	1	1
1	0	0	1
1	0	1	1
1	1	0	1
1	1	1	1

Fig. 12-4. Three-input OR gate with its truth table.

cause either A or B or C, or any combination of these, must be 1, for the output to be 1. Just as in the case of AND gates, OR gates are made with different numbers of inputs. An OR gate with three inputs along with its truth table is shown in Fig. 12-4. Note that F is 0 only when all inputs are 0.

NOT GATES

Unlike the OR and AND gates, the not gate usually has only 1 input and 1 output. Whatever state the input is in, the output is in the opposite state. A schematic and truth table of a NOT gate are shown in Fig. 12-5. Note that the circle at the terminal leading to the output or F line indicates the reversal of a state.

NAND AND NOR GATES

If a NOT gate is placed at the output of an AND gate, the states of the output are reversed from what they would normally be. Thus, the output would be 0 if all inputs were 1 and 1 if any one or more inputs were 0. This is just the opposite to the outputs from an AND gate. This arrangement is the basis of the NAND gate. The schematic representations of this gate along with the truth table are shown in Fig. 12-6. In the first diagram, you see the two gates

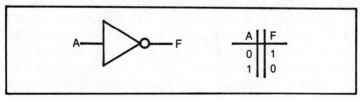

A	F
0	1
1	0

Fig. 12-5. NOT gate with its truth table.

339

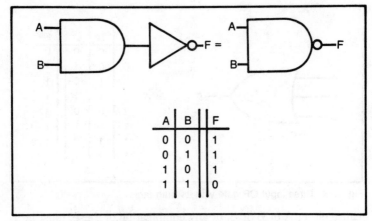

A	B	F
0	0	1
0	1	1
1	0	1
1	1	0

Fig. 12-6. Two-input NAND gate with its truth table.

forming the NAND gate. In the second diagram, you see a representation of the NAND gate as an AND gate with a circle at its output. The entire NOT gate has been replaced by a small circle at the output of the AND gate to show the phase reversal. In the schematic, the entire NOT gate does not have to be drawn to show the conversion to a NAND gate. The circle is sufficient to indicate that the normal output states from the AND gate are reversed.

Similarly, the combination of a NOT gate at the output of an OR gate forms a NOR gate. Here, the outputs are just the opposite of what you would expect from an OR gate by itself. An illustration of this type of gate along with its truth table is shown in Fig. 12-7. Note the circle at the output of the gate to show the reversal of the states.

XOR GATES AND XNOR GATES

Two gates formed from combinations of the three basic gates are the *exclusive OR* or XOR gate and the *equivalence* or XNOR gate. These gates are used frequently in many different digital circuits. Both have only two inputs, A and B, and one output, F.

The output from the XOR gate is 1 when either A or B is 1, but not when both A and B are 1. F is 1 at the output of the XNOR gate when both A and B are either 1 or 0, but not when either input is at 1 and the remaining input is at 0. This is all shown in the truth tables in Fig. 12-8 along with the schematic representations of the two types of gates.

340

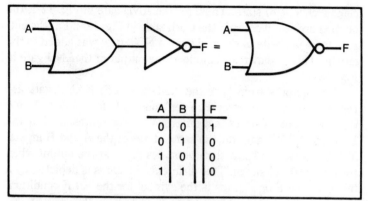

Fig. 12-7. Two-input NOR gate with its truth table.

SYMBOLS AND NOTATIONS

You are now familiar with the basic operation and schematic representation of the NOR, NAND, XOR, and XNOR gates. Circuits may be composed of hundreds and even thousands of different gates. Many gates may be crowded into one integrated-circuit chip.

The NOT Notation

In the depiction of the NAND gate, a circle at the output indi-

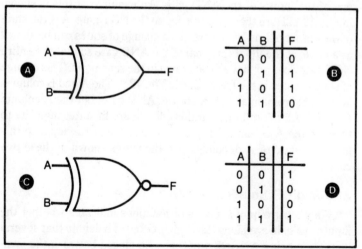

Fig. 12-8. XOR and XNOR gates with truth tables. A and B are the XOR gate schematic and its truth table. C and D are the XNOR gate schematic and its truth table.

cates a change of state. Thus, for the AND gate in Fig. 12-1, the outputs are as indicated in the truth table in Fig. 12-2. For the outputs to be opposite to those shown, a NOT gate was added to the circuit. This is shown as a circle at the output of the AND gate in Fig. 12-6.

The outputs from both the AND gate and NAND gate are represented by the letter F, even though F for the AND gate represents output states that differ from those represented by the F for the NAND gate for the same states of the A and B inputs. There is a way to show that the states differ at the output when the inputs are identical. This difference of states is depicted by a line over the F or $\overline{F}$. This is the symbol for the *not F* condition. So when F = 0, $\overline{F}$ = 1 and when F = 1, $\overline{F}$ = 0.

Look at the NOT gate in Fig. 12-5. The input is labelled A and the output is F. Actually, the same symbol could have been used in both cases. Now the input could have been labelled A and the output is $\overline{F}$ or the input could have been labelled F so that the output is $\overline{A}$. This labelling method shows that the NOT gate provides a change of state. Similarly, the output from the AND gate could have been labelled F, and the output from the NAND gate could have been labeled $\overline{F}$ to show the difference of the states at the outputs of the two devices.

In Fig. 12-6, a NOT gate is shown at the output of the AND gate. The NOT gate could just as easily have been connected to one of the inputs of the AND gate. An example of this is shown in Fig. 12-9. Here the A input is fed to the NOT gate. $\overline{A}$ is the state of the output of this NOT gate. This change of states can be shown as a circle at the input terminal of the AND gate so that the entire NOT gate does not have to be drawn in the schematic. This change of state is noted in the truth table in Fig. 12-9. The A and B columns in the table remain as before for the AND gate, but the $\overline{A}$ column is added. All the numbers in this column are in states opposite to those in the A column. Because $\overline{A}$ and B are at the input to this gate, the state of A depends upon the states shown in these two columns.

AND and OR Notations

Another group of symbols or notations indicates whether the inputs to a gate are being ANDed or ORed. To denote that several inputs are being ANDed, a dot may be placed between the symbols. This dot is usually omitted and the letters representing the input to the gate are placed next to each other. For the AND gate

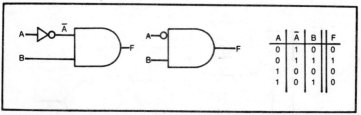

Fig. 12-9. NOT gate at the input to an AND gate with the truth table.

in Fig. 12-1, you can write AB = F, indicating that when A and B are ANDed, the output is F. This is read "A and B equals F." In Fig. 12-3, ABC = F, or A and B and C equals F. For the NAND gate, you can write $\overline{AB}$ = F or "not A and B equals F" because A and B are both ANDed and then NOTed. This differs from the representation in Fig. 12-9 where you have $\overline{A}B$ = F because only A is NOTed before its input reaches the AND gate to be ANDed with B.

Similarly, the OR gate operation can be indicated by simply placing a " + " between the letters representing the inputs. Note that here the " + " does not mean the "plus" used in ordinary arithmetic, but indicates that the gate fed by the inputs, is an OR gate. Thus, A + B + C = F is read as "A or B or C equals F." This is for the OR gate in Fig. 12-4.

CIRCUITS WITH GATES

Using this nomenclature and these symbols, you can represent the various combinations of gates symbolically. For example, consider the combination of gates in Fig. 12-10. These three gates are combined to form the XOR circuit. A and $\overline{B}$ are fed to the input of AND gate #1 so its output is $A\overline{B}$ or A and not B. $\overline{A}$ and B are fed to the input of AND gate #2 so its output is $\overline{A}B$. The two outputs from the AND gates are fed to the OR gate. The output from the OR gate is F = $A\overline{B}$ + $\overline{A}B$. This was noted above in the discussion of this gate. This expression is a way of stating the information stated in the truth table in Fig. 12-8 in the form of an equation. It states that F = 1 when the output is either $A\overline{B}$ or is $\overline{A}B$.

The XNOR circuit has a 1 output only when both A and B are either 0 or 1. Thus, the circuit combination in Fig. 12-11 must be used to produce this type of output. The output from the AND gate #1 is AB and from the AND gate #2 is $\overline{A}\overline{B}$. (Note that the output $\overline{A}\overline{B}$ is not the same as the output for the NAND gate, $\overline{AB}$, where the change of state is at the output. In the gate shown in Fig. 12-11,

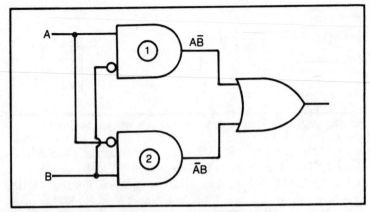

Fig. 12-10. Schematic of gates forming the XOR circuit.

each input is an individually NOTed item, so two individually NOTed items are ANDed. In the case of the NOT gate, two straight inputs are ANDed and the ANDed combination is then NOTed.) When the two outputs from the AND gates are fed to the OR gate, the output is F and is equal to $AB + \overline{AB}$. It can be stated that F = 1 when the output is either AB or $\overline{AB}$. This information is shown in the truth table in Fig. 12-8D.

Numbers can be added through the use of logic gates. A schematic of what is known as a *half-adder* for two binary numbers is shown in Fig. 12-12. Here, the A and B inputs are fed to an XOR gate and to an AND gate. The sum of the bits at the input, S, is at the output of the XOR gate, while the carry number is at

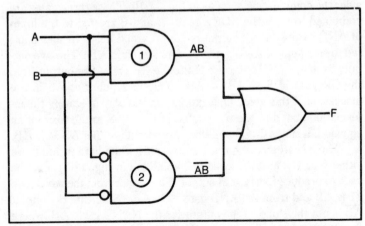

Fig. 12-11. XNOR gate formed from two AND gates and an OR gate.

344

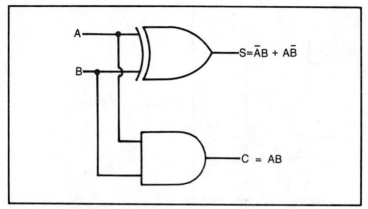

Fig. 12-12. Half-adder circuit.

the output of the AND gate. The term "carry" refers to the more significant bit that can result after two bits are added. If you add 1 and 0, the sum is 1, so there is no carry to a more significant bit. But if you add 1 and 1, the sum is 10 in binary numbers. This is a two-bit number because $1 + 1 = 2$ in decimal notation, and a 2 is a 10 in binary notation. Here, the 1 is the carry bit to the next significant digit.

If A or B but not both inputs is 1, the sum of the two bits is 1 plus 0 or 1. As you can see from the truth table in Fig. 12-8B, when either A or B = 1, the output from the XOR gate is in a 1 state. This is the output at S in the half-adder. When A and B are both 0, the obvious sum is 0. From the truth table, you can see that this is the output of an XOR gate when the two inputs are 0s. Thus, S = 0 when both A and B are 0. Similarly, you will find that the output is zero for the XOR gate when A and B are both 1. You saw that the sum of two 1s is 10, where 0 is the sum and the 1 is the carry. So you can conclude that the XOR gate will produce a sum of 0 when two 1s are applied at the inputs. Now too, S = 0.

Because C is at the output of an AND gate, then C = 1 only when both A and B are 1. Since C is the carry output from the half-adder, and there is a carry of 1 only when adding two 1s, then the AND gate is supplying the carry of 1 under the proper conditions.

FLIP-FLOPS

One of the most used gate combinations is known as a *flip-flop*. An illustration of one of these devices is shown in Fig. 12-13. This type is known as a *J-K flip-flop*. The outputs from other circuits

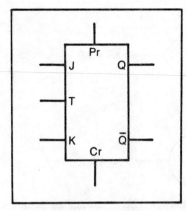

Fig. 12-13. J-K Flip-flop.

or equipment are supplied as inputs to the J and K terminals. Q and $\overline{Q}$ are the outputs from the flip-flop. Note the $\overline{Q}$ at one of these outputs. This indicates that whatever the state of the Q output, the $\overline{Q}$ output is in the opposite state. Q and $\overline{Q}$ are never in the same state.

One of the factors that determines which output is in the 1 state and which is in the 0 state is the state of the J and K inputs. But the output states never change at the instant the J and K inputs are put into specific states. The outputs may change only after a trigger pulse has been applied to the T or clock input. In fact, in many applications a series of regularly spaced pulses is applied to this input.

Observe that there are two more terminals on this flip-flop—Pr and Cr. One or both of these are usually accessible as connections to the IC chip housing this flip-flop. If the flip-flop is not to be affected by the presence of these inputs, these connections must be placed in a 1 or a high state. They have been included as terminals on the chip to enable the user to put the flip-flop into a particular initial state before it starts to fulfill its assigned tasks. If Pr is left in the 1 state and the clear input, Cr, is set to 0 for an instant, the Q output jumps into the 0 state, while the $\overline{Q}$ terminal goes into the 1 state. Should Cr be left in the 1 state and the preset input, Pr, is set to 0 for an instant, Q is put into the 1 state while $\overline{Q}$ flips into the 0 state. After either Pr or Cr is switched to 0 for a moment, the outputs from the flip-flop are in their desired "starting" states. The selected starting states may be necessary in order for the flip-flop to perform its function properly. Both Cr and Pr are then returned to the 1 states so that the J, K, and T inputs can take over. But how does a flip-flop actually work?

Flip-Flop Action

When the J and K inputs are both 1s, the outputs change states with the application of each clock pulse. When used in this way, the arrangement can be referred to as a *Toggle* or *T-type* circuit. But when J and K are both 0s, the states of the outputs never change regardless of the number of clock pulses applied to the T input.

The J-K flip-flop is also used in a *gated latch* or D-type circuit. Here, the J and K inputs must always be in opposite states. This can be accomplished by wiring a NOT gate between the J and K inputs and feeding a 0 or 1 pulse simultaneously to one terminal of the NOT gate and to one input terminal of the flip-flop. The second input terminal of the flip-flop is connected to the output of the NOT gate. It is therefore in a state opposite to that of the first input on the chip. This is shown in Fig. 12-14. When the flip-flop is used in this way, the Q output is in the same state as the J input after the clock pulse, and the $\overline{Q}$ output is in the same state as the K input.

Flip-Flop Applications

A wrist watch or clock is one of the popular digital items that uses a flip-flop. In this as well as in other applications, frequency dividers are necessary. A *frequency divider* takes the frequency of the input signal and divides it down to a frequency that is some portion of what is applied to the circuit. Here is where the flip-flop in the T-type arrangement comes into play. Let us start by assuming that the Q output from the flip-flop is initially in the 0 state. After the first clock pulse has been applied to the flip-flop, the Q output becomes a 1 state. A second clock pulse must be applied

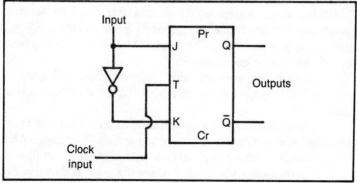

Fig. 12-14. D-type circuit.

347

to bring Q back to 0. Now Q has gone through a complete cycle from 0 to 1 and back to 0. Two clock pulses had to be applied for Q to go through a complete cycle. Therefore, the frequency of the clock pulses was divided by two because two input cycles were needed to produce one output cycle. If this output cycle were applied to the clock input of a second flip-flop, the frequency at the output of the first flip-flop would be divided by two. One-fourth the number of the original pulses from the clock appears at the output of the second flip-flop. The pulse frequency is divided further by two if the output from the second flip-flop is fed to a third flip-flop so that the frequency at the output of this flip-flop is one-eighth the frequency of the original clock pulses. Division by two keeps occurring as long as additional flip-flops are added to the circuit. These exact frequencies are applied to timers to control the accuracy of this display.

Frequency counters are constructed from flip-flops and gates. There are a number of different combinations. These circuits are frequently included in ICs so that the actual details of the various logic systems are not important to the user. A series of pulses is fed to the input of these ICs. The frequency of the pulses is read from the digital display connected to the outputs of ICs. Various frequency readout arrangements are possible.

In one circuit, ten output terminals are available. A bulb is connected to each terminal. When a bulb lights, a number from 0 to 9 is illuminated. When there are eight pulses, for example, the bulb that goes on causes the number 8 to light. When there are four pulses, the bulb goes on so that the number 4 lights, and so on. If there are more than nine pulses, a second set of ten bulbs is used. When one of these bulbs goes on, they cause the next most significant digit to light. This second group of bulbs is usually mounted to the left of the first group. If, for example, there are fourteen pulses, the 4 lights in the first group of bulbs and the 1 lights in the second group of bulbs. The number 14 is thus formed by using two digits—one in each group of bulbs.

There are also readouts using LEDs or LCDs. LCDs are ICs designed to light when a voltage is applied. These use less current than do LED readouts.

In either case, digits are formed from a number of these lights. Up to seven segments are used to form the digits 0 through 9. Only the segments needed to form the desired digit glow. An illustration of ten digits constructed from these segments is shown in Fig. 12-15.

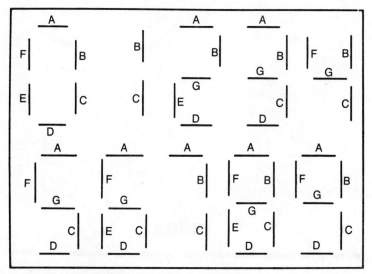

Fig. 12-15. Digital display using seven segments.

Segment A lights only when it is necessary to form digits 0, 2, 3, 5, 6, 7, 8, and 9. This segment lights only when any of these numbers is needed in the display. Similarly, segment E lights only when the numbers 0, 2, 6, or 8 are to be displayed.

What actually happens is that the IC counting circuit feeds the information of the applied frequency to an IC converter. The converter takes the count information in binary form from the counting IC and produces an output. If, for example, the output is from a count of 0101 in binary, or 5 in decimal numbers, voltage from the IC converter is applied only to segments A, F, G, C, and D in the display. When these segments light, the observer sees a 5 to indicate the number counted. Similarly, if 1001 pulses (9 in decimal notation) were counted, voltage is applied to light all segments in the display except for the E, so a 9 may be read on the display.

DIGITAL TECHNIQUES

Instruments using digital circuits have become readily available. A radio or television set may be tuned through the use of some type of digital circuit. Clocks utilize digital readouts such as those described above. Digital circuits are used throughout the entire computer you use at home or in the office. You will also find digital gates in automobiles, high-fidelity audio equipment, alarm systems, and so on. It is a topic worth pursuing further because digital circuits are being applied more and more in everyday life.

Index

Other Bestsellers From TAB